T0220715

Springer Biographies

More information about this series at http://www.springer.com/series/13617

Observatories of the Carnegie Institution for Science Collection (COPC 2806)

Ronald L. Voller

The Muleskinner and the Stars

The Life and Times of Milton La Salle
Humason, Astronomer

 Springer

Ronald L. Voller
New York, NY
USA

Springer Biographies
ISBN 978-1-4939-4382-1   ISBN 978-1-4939-2880-4 (eBook)
DOI 10.1007/978-1-4939-2880-4

Springer New York Heidelberg Dordrecht London

Printed on acid-free paper

Springer Science+Business Media LLC New York is part of Springer Science+Business Media
(www.springer.com)

For Milt

Preface

Milton Humason could have written his own story. His cousin, Tom Humason, an editor at the publisher Harcourt, Brace and Company, wrote him in the fall of 1950 asking Milton if he would write a modern history of the state of astronomical discovery. By then Humason had become a ranking member of the staff at the Carnegie observatories (which combined both Palomar and Mount Wilson observatories) and was considered one of the few people in the world qualified to write such a book. The Big Bang theory was becoming widely accepted as the most plausible explanation for the birth of the known universe, and Humason had driven much of the data behind the theory in an historic twenty-five-year collaboration with his colleague, Edwin Hubble. Together the two men had transformed both public and scientific perception of the visible universe and set the course in astronomy for the next fifty years.

As momentous as the work on the Big Bang concept was, however, it amounted to only a fraction of the work Humason put out during his illustrious and unlikely career. From the publication of his first paper on an obscure comet in 1919 to his last report on the spectra of galaxies in groups and clusters in 1964, Humason published nearly 100 papers with a host of the world's top astronomers and astrophysicists. By 1950, he had worked with almost every member of the sidereal department at Mount Wilson, contributing to the study of star classification, color, magnitude and velocity, stars in clusters and in groups, and novas, supernovas and star populations. His work on galactic structure and universal expansion included his prodigious contribution to the *Hubble Catalogue of Galaxies*, published four years after Hubble's death. In amassing the data for the book, Humason had photographed the spectra of galaxies many times fainter and deeper into space than any before him.

When he wasn't reporting on stellar and galactic evolution, Humason published articles on technological improvements in photography, mirroring, instrumentation and operation. Despite previous notions that he steered clear of the arguments for and against universal expansion, Humason did contribute to the conversation on the topic, which was the center of much public and scientific controversy for decades during the twentieth century. The list of men with whom Humason collaborated during his career reads like a Who's Who of astronomers and astrophysicists— Nicholson, Merrill, Adams, Hubble, Joy, Seares, Baade and Sandage, to name only a few.

In the heyday of the Mount Wilson and Palomar observatories, Humason was, in a very real way, their spiritual leader. As secretary of the Carnegie observatories, Humason was responsible for scheduling observing time on both mountains as well as reading and responding to the preponderance of letters and telegrams that always seemed to find their way to his desk. From inquiries by scientists in varying fields to questions of the existence of God in the heavens, Humason was entrusted to respond on behalf of the observatories and their parent university, the California Institute of Technology, to whom they had recently been linked. Humason was one of the oldest, most respected and best liked of all the members of the observatories and was known around the world as an authority on stellar spectroscopy, instrumentation and deep space photography. In acknowledgement of his achievements, the University of Lund in Sweden awarded Humason an honorary doctorate, and he was also made a member of the Royal Astronomical Society.

Yet despite these achievements, his knowledge of his craft and his standing within the scientific community, Humason declined his cousin's request to offer an account of the state of astronomical discovery at the time. The reason for his refusal to do so ran to the core of who Milton Humason was, as a man and as an astronomer.

Modern accounts of his life and career have created an unbalanced view of Milton Humason. His chroniclers usually paint him as the high school dropout turned muleskinner that rose to scientific prominence as a stellar photographer during the 1920s and 1930s. Lost in these abbreviated and quixotic tales of his exploits was his overriding sense of inferiority that was anything but trivial. Throughout his adult life, Humason's lack of an education beyond grade school undermined him constantly, delaying his hiring at the institution he would eventually help to elevate and creating in him a tense personal struggle between the deferential assistant that he saw himself as and the respected observer and craftsman he sought to become. Well after his accomplishments had earned him the respect and admiration of his peers, his battle against his own sense of inferiority plagued him.

I have deliberately left Humason's better-known attributes out of this introduction to the book for precisely this reason. An honest account of the man cannot be completed without a thorough discussion of both the positive and the negative effects of the attributes that have so endeared him to his admirers. This is by no means to suggest that the attributes that separate him from others in his field don't exist, nor is it meant to discolor the very colorful and charismatic personality that was Milton Humason. He was by all accounts one of the most colorful of all his contemporaries. In fact, the characteristics that have endeared Humason to his legion of fans were the same ones that attracted me to his story.

My association with Humason began while I was vacationing in the Philippines in 2005. Having always been curious about the nature of the stars, I was reading the book *Big Bang* by Simon Singh. In his book Singh offers a rich and thorough history of the developments that led to the Big Bang theory. Around the midway point of the book I came across a brief section 'about Humason who had quit school at the age of fourteen to go to work as a bellboy at the Mount Wilson Hotel. Singh adds that Humason eventually became a muleskinner, helping to haul the supplies and building materials for the observatory buildings and instruments up the steep mountain trail to the peak of Mount Wilson near Pasadena, California. The story goes on to briefly describe Humason's hiring as janitor to the observatory and his subsequent work with Hubble on the expansion problem.

My curiosity piqued, I put down the book and began digging deeper into Humason's background. I soon learned that the details Singh had set out in his book were the only details of Milton Humason's life that anyone seemed to know anything about. Browser searches brought scant traces of his name. A brief biography on Wikipedia offered a few somewhat contradictory details about his life and work. A handful of photographs were scattered among hundreds of photos ranging from astronomers to *The Spirit of St. Louis* to poetry that would have been better left in the drawer. From these few dim clues the trail of Milton Humason vanished into the cold reaches of space-time, leaving me in a quandary as to how to pursue my course. Finally I decided the best way forward was to make my way to Wilson's Peak on the summit of Mount Wilson outside Pasadena, California.

I will never forget my first visit to Mount Wilson in the spring of 2006. As I stood near my rental car in the parking lot of a nearby coffee shop I could see the domes of the 150-foot solar telescope tower and the 60-inch telescope peeking out from the treetops near the summit. Filled with a sense of adventure (that I'm sure many before and after me have felt) I drove to the trailhead and began hiking the old trail up the side of the mountain. I could feel the natural energy of the environment. As I hiked the long and narrow path up the mountain I tried to imagine the excitement of the time when the Mount Wilson Observatory was under construction. The sounds of protesting mules and the young men that drove them up the hill became vivid as I wound my way around the switchbacks and up the steep inclines. I could imagine Milt atop his horse coaxing his charges up the trail, towing thousands of pounds of equipment and supplies behind them.

Once at the observatory, I sat with a small group of people at an abandoned concession stand waiting to take part in a guided tour of the observatory grounds and its instruments. The state of the concession was disheartening. It seemed as though the glory days of the observatory were behind it. Any reservations I had upon reaching the summit, however, were dispelled once I entered the grounds. The observatory is home to a variety of new telescopes and college installations peppered in between the old telescope domes that are still well cared for and maintained.

I had read Helen Wright's book, *Explorer of the Universe*, about the life of George Ellery Hale and could imagine Hale on a visit to the mountain in 1902 (the same year Milton Humason moved to California with his family) clambering up a tall ponderosa pine, dragging behind him a 4″ telescope, to observe the seeing conditions above the mountain floor. I could envision the scene on July 4, 1917, when the 100-inch mirror, standing upright in its huge wooden crate and bound tightly to the back of the old Mack truck, first made its way into the observatory grounds. I stood in the spot outside the 60-inch telescope dome where Andrew Carnegie and George Hale had their picture taken together in March of 1910 and imagined Albert Einstein walking through the grounds chatting with Milt and other astronomers at Mount Wilson in 1931. As I stood on the observing floors of the great telescopes and wandered through the forest that surrounds the observatory I was swept up in the magic of the mountain. Somehow I identified with Milton Humason who, as a boy, fell in love with the mountain and determined that he wanted to be a part of the life there. I was hooked!

My goal in writing this book was to offer the first fully realized account of Milton Humason's life, the people he knew and worked with and the times in which he lived. The era that preceded him, and some of the men who dominated it, were integral to shaping the world Humason was born into near the end of the nineteenth century. For that reason the entire first chapter is dedicated to the changes in the North American physical, social and political landscapes of the time. A small reading list at the end of the book can be referenced for a closer look at any one of the subjects discussed in this brief history. This section is neither meant to establish the complete history of the period nor the people in it, but rather to give the reader a sense of the characters and events that set the stage for Humason's life.

This book is the culmination of nearly ten years of research in cities, libraries, observatories and mountain reserves all over the country in pursuit of the complete picture of the man Harlow Shapley once called "the Renaissance man of Mount Wilson." Humason was, among other things, a practical joker and he enjoyed weaving tales of life on the mountain as much as he did taking the observatory staff for their money in a game of poker. To that end this is as much a story about people as it is about science. From what I was able to glean from interviews with those who knew him, from manuscripts and letters, and from family photos and his published work, I have tried to infuse the book with as much of the substance behind Humason's actions and beliefs as possible. Readers may decide for themselves whether they agree with my assessments or not.

It should be pointed out that I am not a professional astronomer. Much of the science in this book has been gleaned during my research. I am a lifelong avid stargazer and amateur astronomer who, after many years of association with his family and the hunt for information about him, developed a real kinship to Milton Humason. Despite a slight discrepancy in age (he was born seventy-five years before I was) our lives bear some stark similarities. We both had to overcome our own inauspicious start in life, and we both appear to be overachievers. His curiosity and perseverance led Milt to become an expert authority in stellar spectroscopy and the nature of the universe as well as telescope optics, machinery, maintenance and their application on the instruments of his field. I have become an expert on Milton Humason.

The structure of the book runs along a mostly linear timeline. I have made a slight exception to this in the chapter that details Humason's collaboration with

Edwin Hubble, which required some exposition and background to fully understand.

To illustrate the varying degrees of Humason's association with the observatory during its creation, I have chosen to tell the story of the early developments in the construction and design of its instruments and buildings through the eyes of Mount Wilson Observatory founder, George Ellery Hale. In 1895 Hale, then founder of the Kenwood Astrophysical Observatory at his family home in the Hyde Park area of Chicago, Illinois, wrote to his friend James Keeler. Keeler had recently joined Hale on the editorial board of the nascent *Astrophysical Journal,* which had held its first meeting on November 2, 1894, at a 5th Avenue hotel in New York City, and included such scientific luminaries as Albert Michelson, E.C. Pickering and Henry Rowland of Johns Hopkins. Hale was writing Keeler in appreciation of his friend and colleague's recent discovery that Saturn's rings are meteoric in nature. Musing on the seemingly boundless limits of the field of astrophysics Hale wrote: "What a lot of fine fish there are in the sea for the right kind of fishermen."

Milton La Salle Humason was just the right kind of fisherman. Whether fishing the dotted darkness of the California sky or the cool running waters on the West Fork of the San Gabriel River, Humason grew to become a master at delving into the unknown reaches of space and time to extract evidence of the extraordinary world we live in. The life he led and the road he traveled in becoming one of the world's foremost authorities in the field of stellar spectroscopy is equally extraordinary.

His life straddled the late Victorian Era to the early 1970s, an age that began at the end of the Native American epoch in post-Reconstruction America and ended during the Cold War between the United States and the Union of Soviet Socialist Republics. As a boy growing up in Winona, Minnesota, in the 1890s, Humason was influenced as much by the era that preceded his birth as the events that unfolded after it. Legends of the Union victory in the Civil War and Abraham Lincoln's deft negotiation of the war handed down from veterans of the Union Army as well as the ongoing Republican domination in American politics led him to become a steadfast supporter of the party. He remained a staunch Republican throughout his lifetime long after the light had dimmed on Lincoln's era.

Humason was a staff astronomer at the Mount Wilson Observatory from 1922 to 1957. He excelled in the field of stellar spectroscopy, the science of studying the light of distant objects to better understand their structure and evolution. His pioneering work with Edwin Hubble in the late 1920s and 1930s led to discoveries that helped to set the foundation for our understanding of the world we live in, opening our eyes along the way to the seemingly boundless nature of the known universe.

Humason was especially gifted at gathering data for very faint objects. He spent more than forty years trolling the universe for signs of light from objects many thousands of times fainter than can be seen with the naked eye, often testing the limits of the most powerful instruments of his day as well as his own skill and endurance. At the telescope his shrewd thinking, deftness and unwavering dedication helped unravel some of the great mysteries of science. When he wasn't working at the observatory his favorite pastime was fishing the waters of the West Fork of the San Gabriel River.

Carl Sagan once wrote, "The surface of the earth is the shore of the cosmic ocean." It is a vast ocean, too, some 13.7 billion light years distant. We live on a lonely planetary outpost on the cosmic landscape we call Earth, revolving around a 4.6 billion year old class G2V yellow star (one of as many as 400 billion or so stars in our galaxy) on the Orion arm of a barred spiral galaxy known as the Milky Way, about 27,000 light years from its center. Our Solar System, like many across the cosmos, is enriched through the constant production of basic elements such as helium, nitrogen and oxygen by our parent star, and the occasional and very volatile explosions of stars in supernovas, which provide us with the other heavier elements that make up our home planet. Our home galaxy is roughly 100,000 light years across and is part of a universe that contains hundreds of billions of galaxies, each containing hundreds of billions of stars. A section of the sky seen through a straw 8-feet long might reveal as many as 10,000 galaxies if viewed with a powerful enough telescope.

That is the world we live in. What we know of it may be incredibly little in the grand scheme of things. What knowledge we do have, however, we owe in part to the work of Milton L. Humason, one of the most charismatic men of science, whose humble nature, skill and perseverance helped open our eyes to the vast and powerful nature of the universe and set the standard for generations of astronomical research.

Acknowledgments

I have to thank the three stars in my Humason constellation: Mary Brown, president of the Los Angeles Astronomical Society, for introducing me to the mountain and to Don Nicholson; to Don, the son of Seth Nicholson and former director of the Mount Wilson Observatory Association, for introducing me to Milt and to Dan Lewis; and to Dan, Chief Curator of Manuscripts at the Huntington Library for his patience, selflessness and guidance during my research. Also Dave Eicher, editor at Astronomy Magazine and Maury Solomon at Springer for their trust and support. John Grula, head librarian at the Mount Wilson Observatory offices and Luisa Haddad, librarian at the University of California at Santa Cruz were very helpful during my brief visits to their institutions. My warmest appreciation also goes out to Harry Wing Humason, Milton Humason's nephew, for his input on the family history, and to Laurie DeJong, CEO at LDJ Productions for allowing me the freedom to continue to explore. Finally, and most importantly, I would like to thank Ann Humason Bernt, without whom this book would not have been possible. In addition to these few, there were countless contributions made to the creation of this book by people at various libraries and institutions across the country, and I am grateful for them all.

Contents

About the Author

Ronald L. Voller is an author, amateur astronomer, artist, historian, strategist and creative director who lives and works in New York City. He has been a Reader at the Huntington Library in Pasadena, California, for the past ten years. Mr. Voller is also a board member of several charitable foundations and regularly contributes to the education and economic growth of communities around the world. When he isn't working, Mr. Voller enjoys reading, writing and playing music, sculpting, sports and is an avid outdoorsman.

List of Figures

Prologue

Abstract

The events that transpired in the last forty years of the nineteenth century set the stage for the incredible industrial and technological growth during the first half of the twentieth century. From the desolation of Native American peoples in the Central Plains, to the completion of the transcontinental railroad, the most monumental moments and the men who shaped them contributed greatly to the American Century that was to follow. This chapter explores the events that transpired that would directly affect Milton Humason and the world he lived and worked in.

Pioneers, Innovators and Capitalists (1859–1902)

The four decades leading up to the turn of the twentieth century, when a young Milton Humason was just entering his formative years, were marked by events that would set the stage for the American Epoch. Civil and natural rights were laid bare on the world stage, the beginning of corporate industrialization began destabilizing the balance between rich and poor, and great feats of engineering and advances in science and technology made communication and travel across great distances increasingly easy and efficient. Many of these events, and the characters involved in them, were instrumental in the life of Milton Humason, who grew up in a time when the United States was beginning to emerge from the turmoil of the Victorian Era into its role as a world power.

In 1859 the United States of America stood on the edge of unprecedented changes to its geographic, social and political landscapes. The country was divided in these ways along two well defined lines of demarkation east and west as well as north and south.

The east-west boundary lay along the 95th meridian (on the western borders of states from Minnesota south through Louisiana), dividing the civilized east from the frontier west of the country. Life in the East was becoming increasingly industrialized with sprawling cities, locomotive transportation and modern communication by telegraph. The first zoological foundation had been established in Philadelphia and intercollegiate baseball was being played between Amherst and Williams College. Buildings, bridges and even ships were being built of iron instead of wood and brick and mortar. Law and order was taking hold in cities large and small, where increasingly streets were paved in stone and lined with sidewalks for pedestrians to move freely away from the bustle of the horse and buggy traffic on the roadways.

The West, on the other hand, was still a rugged and treacherous place, where Native American tribes still roamed the Great Plains and frontier law was the order of the day. Travel west of Omaha in those days meant risking life and limb for those daring enough to try to build a new life beyond the limits of lawful society. Still, increasing numbers of Americans were braving the frontier in search of a small stretch of land to call their own. Texas, California and Oregon were the only recognizable states in the West, and there were untold stories of glorious wealth and grandeur to be gained for the adventurous soul. However, many who ventured forth to find their fortune died along the trail from starvation, disease, run-ins with Native Americans or armed bandits, often before they reached the Continental Divide.

The north-south boundary presented a different sort of divide, one that threatened the very existence of the nation. Slavery, at issue since before the Founding Fathers signed their names to the Constitution of the United States and one many feared could only be resolved through bloodshed, was beginning to boil over into increased conflict. The Missouri Compromise of 1820 had divided the country in half along the 36th parallel, which cuts through the southern border of Missouri, east of the Mississippi River. Under this law slavery was made illegal in the former Louisiana Territory north of the 36' 30" parallel and west of the Mississippi River except in Missouri, where it was made legal as part of the deal struck between pro-slavery and antislavery factions in Congress. (In 1820, Spain still owned the southern states of Texas, New Mexico, Arizona, Utah, Nevada and California.) Despite the compromise slavery remained a hotbed for dissent on both sides of the issue.

In 1854, Congress passed the Kansas-Nebraska Act, giving white settlers in the new territories of Kansas and Nebraska the power to choose whether they wanted to own slaves through popular sovereignty. The new law effectively nullified the Missouri Compromise and set the wheels of civil unrest in motion.

In its ruling on the Dred Scott versus Sanford case of 1857, the Supreme Court declared that Congress did not have the authority to prohibit slavery in the territories and that the Missouri Compromise was unconstitutional. The decision, which said that Scott, a former slave from the South who was suing for his freedom, whether free or not, could not be an American citizen and therefore had no standing to sue in federal court, weakened the court and subjected it to public opprobrium from the left, which accused it of racism, trading basic human rights in favor states' rights.

The Supreme Court ruling in the Dred Scott case further fomented violence between pro- and antislavery factions. In 1859, a militant abolitionist named John Brown led an assault on the U.S. armory at Harper's Ferry, West Virginia. Brown's men seized the armory, intending to use its arsenal to arm local slaves who would then fight for their freedom. Brown had gained national attention at the Pottawattamie Massacre of 1856 in which five pro-slavery supporters were killed during a campaign to step up opposition to the institution of slavery through violence. Brown believed that the endless talk from antislavery supporters had proved impotent and that only violence against slave owners and supporters of slavery would rid the country of what he believed was a sin against God's will. Brown saw himself as the instrument of that will. His efforts fell short, however, as he and his forces were captured at Harper's Ferry by U.S. Marines led by Robert E. Lee. Brown was later executed by hanging in the state of Virginia, after being convicted of conspiracy, murder and treason.

It was against this backdrop, in 1859, that a man named Benjamin Davis Wilson bought a vast area of land near the pueblo of Los Angeles, California, land now known as Pasadena and Altadena. The purchase included a large area of the San Gabriel Mountains to the east. An old trail, cut by Native Americans living in these hills, led to the top of the 5700-foot mountain above Wilson's home in the valley. From the peak Wilson loved to stand and gaze down on his land and the fledgling town of Los Angeles stretching out through the valley toward the Pacific Ocean. The summit of the mountain eventually became known as Wilson's Peak.

Wilson had come to southern California along the Old Spanish Trail from Santa Fe, New Mexico, in 1841. During this era, anyone homesteading between the Mississippi River and the western slope of the mountain range of California was taking a hard road. These early pioneers of the old West were a rugged, highly spirited and resilient people. Wilson, a veteran and former prisoner of the Mexican-American War who became one of the handful of brave mountain men who worked as a fur trapper and traded with the Native Americans, was as tough as they come. Six feet tall and weighing 200 pounds, with a steely eyed stare and pronounced brow, Wilson cut an impressive and handsome figure that served him well in both the halls of government in Los Angeles and the kill or be killed world of frontier life. At the time, in California, one was usually not far from the other.

On one occasion in the 1840s, Wilson tracked a grizzly bear into the woods near his home on Jurupa Ranch, where there were some 300 head of cattle. At the time California was part of Mexico, and Wilson served as alcalde (mayor) of the district he lived in. The bear had been taking Wilson's livestock and, to protect his interests, Wilson decided the bear had to be killed. He set out with a hunting party to find and kill the beast. But the hunter soon became the hunted. The bear stalked the party and then ambushed the men, mauling Wilson before the bear was driven off by his dogs. Wilson was taken back to his ranch, where he slowly recovered from wounds to his hips and arms. Meanwhile the bear resumed taking his cattle, which left the ailing rancher fuming.

Once healed, Wilson took up the hunt again. This time a cow carcass was used to lure the bear under a sycamore tree, where Wilson and his associate waited with guns loaded and ready. When the bear appeared they each managed to shoot it once, but this only served to anger the bruin. Bellowing angrily the bear tried to climb the tree after Wilson and his fellow hunter but was hampered by its wounds. It soon tired and ran off, with Wilson and his men in hot pursuit. The party found the animal lingering in a mud bog nursing itself back to health some time later. Wilson dismounted and took aim, but the bear started and leaped on him, mauling him again before succumbing to a hail of gun fire. Wilson had finally prevailed, but the bear had left its mark. The wounds he suffered would leave Wilson scarred for life.

In the years leading up to the California Gold Rush of 1849, although he had no formal education, Wilson became well known for his shrewd business sense. With his young wife, Ramona, by his side he began buying land for his large herd of cattle. He made friends in the local government, and through his connections was able to increase his real estate interests. As California entered the union in 1847 Wilson was in a unique position to capitalize on the opportunities that would follow. He ventured into merchandising and continued to buy up property all around the area. Following the Gold Rush, his interests made him a fortune as new wealth poured down from the mountains. By the mid-1850s, he had become one of the wealthiest men in California. He served as mayor and was elected to the state senate in 1856 representing Los Angeles county.

Don Benito, as he was known, continued to expand his estate, becoming the nation's leading vintner through the middle of the 1860s. After the Civil War, Wilson donated a large plot of land, used by the Union Army (Camp Drum) during the war, for a college. Wilson College later became the University of Southern California. Benjamin D. Wilson died in 1878, one of the most respected men of his generation. He was the grandfather of General George S. Patton, the famous commander of Allied Forces during World War II. Wilson's Peak was later renamed Mount Wilson and, at the turn of the twentieth century, became the home of the Mount Wilson Observatory.

* * * * * * * * * * * * * * * * * **

While Benjamin Wilson was buying up land in the Los Angeles area in 1859, another man named Theodore Judah was hiking through the Sierra Nevada range 400 miles north. Judah was plotting, mapping and surveying the area in search of a suitable path for a railroad that would connect the continent from east to west. His preferred path led from San Francisco into the mountains through Donner Pass, so named for the ill-fated Donner Party expedition of 1846–47. The terrain was rugged and the weather unpredictable, as members of the Donner expedition had learned to their dismay and peril, losing more than half of their party to exposure and starvation after being trapped at the pass by snow and cold. The harrowing circumstances had caused those who remained to resort to cannibalism to survive. Nevertheless, Judah was certain he could open an acceptable route through the area and into the Northern Plains.

When James Buchanan (the last democrat elected to the presidency of the United States until Grover Cleveland in 1885) handed over the keys to the Oval Office to Abraham Lincoln in 1861, Lincoln took up the issue of a transcontinental railroad that had been started during Buchanan's administration. The following year, with the Civil War raging across the southern United States, Lincoln commissioned the Union Pacific and Central Pacific Railroad companies to build the railroad, and work soon began on the grade for the line, with the UP working its way west from Omaha, Nebraska, and the CP working east from San Francisco.

Judah, a thirty-six year-old civil engineer with a genius for prospecting, designing and building railways over unforgiving terrain, was given the unenviable task of managing the building process. Intelligent and earnest, with melancholy eyes that peered out between a head of thick, wavy brown hair and a bushy beard, Judah suffered countless setbacks in his role as liaison between the government and private interests connected with the railroad. Through it all his ambition for the creation of the railroad, one of the greatest feats of engineering in history, remained steadfast. His disdain for the corrupt men who were reaping fortunes from the building of the railroad created a deep and lasting depression in Judah. With the government focused on the war effort there were few, if any, resources at his disposal to reign in the corruption. Still, he tried to put all this aside and focus on the construction of his masterpiece.

The obvious terminus for the eastern portion of the railroad was near Omaha, Nebraska. All the roads in and out of Chicago were leading there, and Judah determined that the grade across the Northern Plains was best suited to the creation of a train line to California. Relatively speaking, the eastern portion of the line was easy. The tricky part was getting through the Sierra Nevada Mountain range from the California coastline. A telegraph running from Missouri to San Francisco was completed in 1861, making communication easier between Judah and railroad executives eagerly awaiting updates on progress. The route would run through Donner Pass, on a saddle between the higher elevations, winding its way up and down through the steep mountains. The path through the hills was cut in long sweeping turns designed to maintain the minimum grade acceptable for running the most modern locomotives through the range.

Judah had made a full set of surveys, maps, and cost projections during his reconnaissance of the area in 1859. He used these to woo a group of railroad men led by the California governor, Leland Stanford, and railroad magnate Collis P. Huntington, into creating a company that would become the Central Pacific Railroad. Huntington and Stanford formed a consortium of businessmen known as the Big Four, with Stanford acting as president and Huntington as vice-president. Charlie Crocker, co-founder of the CP, was in charge of construction, and Mark Hopkins, a merchandiser who had made a fortune selling goods during the 1849 California Gold Rush, was treasurer.

Almost from the start there were problems in the management of the project. Judah was suspicious of Huntington and accused him and his partners of siphoning off money on Central Pacific stocks owned by companies they had hired to do the work. The CP would award contracts to dummy individuals through Charlie

Crocker's Contract and Finance Company. Once the contracts were awarded, the CP would write a check to Crocker's company, which was used to buy CP stocks and bonds. These were then sold on the open market with the Big Four collecting dividends on the surplus revenue from the deals. Judah and everyone with an interest in the railroad knew what Crocker and Huntington were up to but were powerless to stop it.

Huntington used the political and economic might of the Big Four to strong arm Judah, telling him if he didn't like the way things were being done at the CP that he was welcome to buy him and the others out. To Judah, the railroad was his, and he was doing his best to ensure it was being built properly. Huntington and his partners were simply stealing from the construction of the railroad to line their pockets. Lacking a fortune of his own Judah sought the financial support of railroad and shipping magnate Cornelius Vanderbilt in New York City. Vanderbilt had shown interest in buying out the Big Four, and Judah set out with his wife in early 1863 to seek Vanderbilt's endorsement and financial support in trying to do just that. In a cruel twist of fate he contracted yellow fever while crossing the Isthmus of Panama and died in New York City a short time later, having never met with Vanderbilt.

Thus Theodore Judah never saw the successful completion of the transcontinental railroad, nor did he win his battle against Huntington and the other Central Pacific owners. He was, however, highly instrumental in the design and building of the railroad, which was completed in Promontory, Utah, on May 10, 1869, ushering in a new era of cross-country travel that would in the years that followed help bring civilization to the West.

The transcontinental railroad was a marvel of modern engineering. A trip that just the year before meant months of harrowing travel across Native American lands by horse and wagon could now be completed in a week. By 1880, the railroad network in the United States nearly doubled with rail lines reaching every territory. The relative ease of crossing the great expanses of the country led to increases in immigration, agriculture and manufacturing throughout the West. By 1891, the year Milton Humason was born, the total network reached almost 164,000 miles of track and included several major cross-country trunk lines from north to south. In the West, the network had gone from 2175 miles in 1860 to over 72,000 miles of track by 1890, and the population of the western portion of the United States had grown four fold in that time.

* * * * * * * * * * * * * * * **

It is hard to imagine the effect the transcontinental railroad would have had on someone like Sitting Bull, the great tribal shaman and chief of the Hunkpapa clan of the Lakota Sioux. For hundreds of years, well into the nineteenth century, his people reigned over the Northern Plains of North America, migrating along with the enormous herds of buffalo that roamed throughout the northern and central plains. Sitting on his horse, high above the North Platte River Valley in what is modern day Nebraska, watching as the road was being built, the sense of awe, hopelessness and urgency for his people would have been overwhelming. Hundreds of men concentrated around two narrow sections of iron rails, that stretched to the horizon to the East, were pounding spikes into large wooden ties to secure the tracks. Ahead of them, still others laid more ties into the freshly finished stoney grade of the railway.

Just behind the workers to the east, the shrill howl of a giant iron horse broke the silence of the early morning. The great beast sat spewing mushrooming tufts of gray smoke and fire from the stack atop its round belly as the army of workers toiled swinging sledge hammers, carrying heavy road ties and shoveling and raking stones for the road. The distant tinny ping of hammers on spikes must have rung in Sitting Bull's ear like slow beats from an enemy's drum, as the invaders incessant rhythmic dance went on.

Sitting Bull had seen these iron horses before on his rides along the eastern border of the Dakota Territory. But until then, no attempt had been made to build a railway west of the Missouri River. This incursion into Lakota lands would have to be avenged, and Sitting Bull was no stranger to fighting for his people. He had been leading them in skirmishes against U.S. army installations for years in support of other Sioux chiefs such as Red Cloud, whose war against the United States raged through the latter half of the 1860s. Although the government sought a treaty to end hostilities with the great Oglala Sioux chief (signed by Red Cloud at Fort Laramie in western Nebraska on April 29, 1868), Sitting Bull refused to sign the bill and vowed to continue attacks on army strongholds. But his efforts, like so many other Native American tribal leaders, were unable to stem the tide of American settlements in the west.

In 1850, Native Americans roamed with relative impunity over most of the United States west of the 97th meridian. By 1870, that territory had been reduced to a quarter of its size, sending refugee tribes into reservations where they invariably suffered, falling victim to disease and depression. Although large as territorial land grants were concerned, the land the Lakota Sioux once called home was but a fraction of its former grandeur.

For Native Americans in the West, the next twenty years would be punctuated by a few pyrrhic victories that were inevitably superseded by loss, anguish, ignominy and helpless despair. Some of the greatest warriors of any age were eventually subdued by the incursion of western civilization. A few, like Quanah Parker, the great Comanche chief, managed to make due in their new life on secluded territorial farmland. As a warrior Parker, the son of a white woman and whose father had been chief of their tribe, was among the fiercest of his day. Although other tribal leaders of the Great Plains were signing treaty after treaty in the hopes of saving a portion of their land for themselves and their posterity, Parker never put pen to paper. Once resigned to his people's fate, however, Parker led them to a reservation in southern Oklahoma and became a successful rancher and a founder of the Native American Church. A fierce public advocate for the Native American cause, Parker played a central role in their assimilation until his death, from heart failure, in 1911.

Sitting Bull met with a vastly different fate. A powerful and admired spiritual leader, he had once foretold of a great victory against the U.S. soldiers in a great battle. On June 25, 1876, riding alongside Crazy Horse, Sitting Bull led his people to defeat Lt. Colonel George A. Custer and the 7th Cavalry at the Battle of Little Bighorn. The victory was complete and decisive and increased Sitting Bull's already immense standing within the Native American community. Shortly after the Battle of Little Bighorn, however, he and his followers were driven over the Canadian border by U.S. forces. He remained there until the cold Canadian winters and the threat of starvation forced him to leave Canada and head back into the United States, where he and his tribe surrendered to U.S. forces at the Standing Rock Agency.

Once back on the reservation, Sitting Bull took up his people's cause, advocating for Native American rights in government. In 1885, he spent some time with Buffalo Bill Cody's Wild West Show, which catapulted him to fame and fortune (most of which he gave to homeless people). After months on the tour Sitting Bull had had enough and returned to the reservation, where he picked up the political fight for his people. Five years later, on December 15, 1890, he was confronted at his home by Indian Police, sent to apprehend him for fear he would leave the agency with members of the Ghost Dance movement, who believed the spirits would grant them peace, unity and prosperity separate from the anglo world. A scuffle ensued, and Sitting Bull was shot by one of the officers and died later that day. In life as in death, Sitting Bull embodied the spirit of the Native American people. The Massacre at Wounded Knee two weeks later effectively ended the era of open and free Native American culture in North America.

* * * * * * * * * * * * * * * * * **

In the early hours after dawn on a cool and damp October morning in 1871, William Ellery Hale walked to the corner of Washington and State Street in downtown Chicago, Illinois. Standing on the corner where the Hale Building, a manufacturing plant that had been constructed to house his elevator business, stood just two days before, Hale could hardly believe his eyes. The plant had been reduced to a smoldering mass of rubble, one of many buildings destroyed in one of the worst fire disasters the country had ever seen.

Born in Bedford, Massachusetts, in 1836, the son of a congregational minister who had moved the family to Wisconsin to enter the paper business shortly after William's birth, the adult William had made his way to Chicago shortly before the start of the Civil War. He struggled to make it in the real estate business and later met Mary Scranton Browne. The two were married on Christmas Day, 1862, with the war raging, and moved into a boarding house on Indiana Avenue. In 1863 they had a son, Edward, who died after seven months. Two years later a second child, Caroline Scranton, died weeks after her birth. After the war, as Reconstruction began, William's prospects in real estate improved, and he and Mary moved into a more comfortable home at 236 North LaSalle Street, where their son George Ellery Hale was born on June 29, 1868. George was the first of three Hale children who would survive to adulthood.

Through the mid to late nineteenth century, Chicago became the industrial center of the United States. The advent of a cross-country rail system together with

Chicago's location on Lake Michigan and near the western frontier of the country, made the city an ideal hub for transportation of goods and supplies in every direction. In a rush to establish itself as the fastest-growing city in the region, Chicago's downtown underwent historic growth in the years following the war. Although the long-winded and blustery politicians (for which the city got its nickname of the Windy City) bragged of the indestructible nature of the new buildings with their fireproof facades, the jerry-built structures erected within the Loop District were little more than a vast array of tinder boxes. An article in the *Tribune* eerily foreshadowed the coming doom: "Chicago is a city of everlasting shingles, sham, veneer, stucco and putty."

It was within one of these very structures that the Hale Building was completed in 1870. The new manufacturing facility had housed the Hale Water Counterbalance Elevator Company. William Hale had patented this new type of hydraulic elevator the year before and, through his contacts in the banking and real estate industries, had located the site and funding for the building and its machinery. With his new elevator design and a plant to build it in, Hale was poised to become a powerful player in the development of new high rise buildings being planned for the heart of downtown. With the future looking more secure, Hale decided to move the family south of downtown to nearby Hyde Park. This uncanny stroke of luck spared them the fate now suffered by many of their friends.

At around 9 p.m. Sunday night, on October 8, 1871, fire broke out in a barn behind Patrick O'Leary's cottage at 137 De Koven Street. Fueled by high winds, the blaze quickly spread east, jumping the south branch of the Chicago River on its way to the shoreline of Lake Michigan before raging northward, consuming everything in its path. Feeding on the intense heat and fiery winds, the inferno burned through an area two-thirds of a mile wide and four miles long. Massive pieces of flaming timber and ash flew overhead, landing on rooftops, setting ablaze building after building as the fire ate its way through the northern streets toward Lincoln Park. Later, shocked and dismayed, citizens would tell of the chaos and hysteria that rained on the streets as the fierce wind propelled the flames high into the night sky, causing residents to flee their homes to seek shelter near the lake-shore. Many were overwhelmed by the smoke and fire or crushed under wagon or horse. In the rush to preserve what possessions they could, some residents quickly buried their things underground behind their homes and businesses. From the safety of their home the Hales watched helplessly as the city erupted in a ball of fire. The fire left a lasting impression that 3-year-old George would remember many years later: "It was one of those curious flashes which strangely linger long after the vast throng of less tenacious illuminations have sunk into obscurity."

Finally a steady rain started just after midnight on the 10th, extinguishing the fire and leaving the city in a state of utter desolation. The toll from the fire was staggering. Three hundred people were dead, 100,000 had lost their homes and more than $200 million in property had been destroyed.

On his way into the city that morning Hale witnessed firsthand the toll of human suffering. Tens of thousands of rain-drenched and beleaguered townspeople were huddled along the lakefront, their hands full of the few possessions they could physically carry out of harm's way. Everything imaginable was strewn along the shore—pianos and trunks, drawers and hope chests, paintings, lamps, countless papers. Alongside these sodden artifacts sat the beleaguered people of Chicago. They had dragged, hauled and pushed their family treasures to the shore only to see them burned by falling debris as the raging inferno strafed the shoreline. After nearly two days of terror, the drenching rain had moved into the area, soaking their flagging bodies and spirits as it put out the fire.

Within the city the specter turned even more gruesome. The lifeless bodies of people and animals lay indiscriminately in roads and alleys. The streets were thick with smoldering debris. Once familiar buildings and locations were virtually indistinguishable, save for a few facades that had, by chance, been left standing.

The sky still rained down with bits of falling ash as Hale stood helplessly surveying the ashen ruins of his factory. The wasteland of the downtown district left in the fire's wake mirrored the despair and desolation that now invaded him. With the family's fortunes up in flames Hale was overcome by emotion.

Shortly after his arrival at the site, however, Hale was met by his friend Lyman J. Gage, of the First National Bank. For a while the two men traded wide-eyed stories of the misery they had witnessed. In time, the conversation turned to revitalizing and rebuilding the city. The bank held the mortgages on both Hale's home and business, and to ease his friend's fears, Gage promised to extend credit to Hale so that he could rebuild.

In the years that followed, as Chicago rose from the ashes and Hale elevators were installed throughout the city's new skyscrapers, William Hale's fortune from his real estate and elevator businesses grew. Along with Elisha Otis, William Hale became one of the few men credited with helping to build the world's great industrial cities in the latter half of the century. Hale elevators were installed throughout the United States and Europe. William Hale's real estate deals included the building of the Rookery Building in 1886, which was designed by Burnham and Root Architects. The building established an inventive new process, developed by John Root, that used a steel skeletal structure that allowed for a more open style to the inner spaces of the building. In another venture, Hale bought the Reliance Building that, at the time, housed the First National Bank. Bank officials were reticent to relinquish their lease, so Hale had the top four floors of the building raised using jacks and then hired Burnham and Root to design and rebuild the bottom floor and basement in the new style created by the firm. In time Daniel Burnham and William Hale became friends, and Burnham became a mentor to the elder Hale's son, George.

As his son's interest in the stars grew, William Hale did everything he could to provide George with the instruments and education he needed to excel at his chosen field. In the late 1880s, at George's urging, Hale built an observatory on the grounds of the family's estate in Hyde Park, and furnished it with a 12-inch telescope the boy had asked for to pursue astronomical matters. When George declared an interest in studying astronomy, it was Daniel Burnham who recommended him to Edward Pickering for the graduate astronomy program at M.I.T. William Hale's greatest gift to his son came in 1894 in the form of a 60-inch glass disc that George

Hale intended to use as the primary mirror for what would become the largest telescope in the world. The telescope went into service ten years after William Hale's death in 1898.

George Hale and Daniel Burnham would continue to work together through the first twenty years of the Twentieth Century. It was Burnham who designed the dome features of the 100-inch telescope on Mount Wilson. William Hale's greatest legacy to his son was in his boundless enthusiasm and excellence in the field of engineering, both attributes that George Hale inherited from his father and both of which he used with the goal of bettering the world and increasing our knowledge of the physical universe.

* * * * * * * * * * * * * * * * * **

James Lick wasn't used to receiving guests unannounced, let alone while he was still in bed. In his weakened condition, though, he wasn't making much fuss about it, especially after he learned it was his old friend George Davidson calling. With the help of his nurse, Lick propped himself up in bed inside the hotel that bore his name, as his friend entered the room. Davidson was president of the California Academy of Sciences and the man most responsible for convincing James Lick to build an astronomical observatory on top of Mount Hamilton in nearby Santa Clara. When completed, the facility would become the world's first fully functioning year-round mountaintop observatory and house the world's greatest telescope.

The date was August 25, 1876, and Davidson had come to wish the wealthy philanthropist a happy eightieth birthday. Anxious for any news regarding the institution that would define his legacy, Lick asked him for a progress report. Construction of the road to the top of the 4200-foot peak was underway and going smoothly. Lick had made his endowment for the state of the art facility only if the county would provide a suitable road to the top of Mount Hamilton, a demand the county quickly met in view of the prestige Lick's gift would bring to the area. On hearing the news of the road construction, Lick breathed a sigh of relief. After years of toil, a lifetime spent amassing his vast fortune and much hand-wringing over the realization of his legacy it seemed clear that his wishes would finally be granted. The gray beard on his chiseled face and his stern expression betrayed the eccentricities and relentless ambition that resulted from an unfortunate episode in his youth.

Born in Stumpstown (Fredricksburg), Pennsylvania, in 1796, James Lick had been taught the fine craft of woodworking by his father. In time the younger Lick grew to surpass even his father's excellence in the craft. He remained an apprentice to his father until the age of 21, when he met Barbara Snavely, the miller's daughter. The two fell in love, and soon his beloved was with child. Wishing to do the right thing, Lick asked the miller for his daughter's hand in marriage. But the miller refused, telling Lick that his daughter deserved better than an apprentice in a wood shop. When Lick had as fine a mill as his, the older man said, he could have his daughter's hand, but not before. In a fit of rage Lick yelled as he slammed the door, "Someday I will own a mill that will make yours look like a pigsty!"

Beside himself James Lick soon moved to New York City, where he took up his craft and soon made a name for himself creating fine pianos. When he realized many of the pianos he was building were being shipped to Argentina for sale there, he decided to move to the seller's market and boarded a ship for the South American country. After several years learning the native culture and language Lick's business in pianos was thriving. In 1832, Lick decided to return to Pennsylvania to ask for Barbara Snavely's hand. He was now a rich man and carried with him $40,000 in furs that he planned to sell during his visit. Upon arriving, however, Lick received some unwelcome news; his father had died the year before, and Barbara Snavely had married and had gone into hiding with their fourteen-year-old son, John. After not hearing from Lick for a decade and a half she had made a life for herself, and on his return to the city she could not face him. After two weeks searching in vain for his beloved Barbara, Lick left the city and headed back to Buenos Aires. The event had soured him, and he vowed never to open his heart to a woman again as long as he lived.

For the next few years Lick moved first to Chile and then Lima, Peru, where he again used his expertise in the building of pianos to prosper. It was there, between the years 1835 and 1848, that Lick befriended a young chocolatier named Domingo Ghirardelli. As the Mexican-American War was ending Lick decided to sell his piano business and move to California. Sure the United States would win the war and annex the state from Mexico, Lick wanted to try his luck on the newly minted American soil on the shores of the Pacific Ocean. He arrived in the port of San Francisco on January 7, 1848, a month before the war with Mexico officially ended, with $30,000 in Peruvian gold, a work bench and tools, and 600 pounds of chocolate from his friend, Ghirardelli. After selling the sweets in short order, Lick wrote to his friend in Peru and told him he should move his business to San Francisco, which he did, establishing the Ghirardelli Chocolate Company in 1852. To this day the company is a landmark in San Francisco.

Lick wasted no time reinvesting his fortune in the new American state of California. His timing in business was as good as his timing in love was bad. On January 24, 1848, John W. Marshall found gold at Sutter's Mill, the beginning of what would become known as the California Gold Rush. Between February and March of that year, James Lick bought up 37 lots of land in the San Francisco area. As the population of the sleepy hamlet went from 1000 to more than 20,000 in two years, Lick's investment in real estate turned a tidy profit. By 1850 he had become one of the richest men in California.

Never forgetting his vendetta to the man who had refused his marriage to Barbara Snavely so many years before, Lick soon began work on a mill so lavishly furnished that when it was completed in 1855 he had it photographed and sent the photos to Stumpstown. There is no end to the lustful creativity and zeal of a spiteful man.

As the years progressed through the end of the 1870s Lick's fame for his contributions to the building of San Francisco shared equal billing with his reputation as an eccentric. His 24-room mansion in Santa Clara remained largely unfurnished and lacked drapes into the early 1870s. Instead of a providing a suitable bed, Lick preferred to sleep on a mattress sprawled across a piano a friend had made for him. During the Civil War he built the Lick House in San Francisco, the most expensive and finest hotel in the West, where he had taken up residence after a stroke he suffered in the spring of 1873 left him disabled.

Over the course of the next several years, Lick spent his time in pursuit of his legacy, devising some unusual and unwelcome ways to divest of his fortune. His first idea was to have enormous sculptures of he and his parents erected along the San Francisco coastline, so that they could be seen for miles offshore. His handlers convinced him, however, that these would likely be destroyed should the United States ever come under naval attack offshore. Next, Lick decided to build a pyramid that would eclipse the Great Pyramids in Egypt in size and scope. It was during this time that his friend George Davidson convinced him to forego the pyramid scheme for an observatory that would cement his legacy forevermore in the name of scientific research.

It had taken some effort to convince Lick that the observatory should not be built in the spot where the pyramid was to go but high atop Mount Hamilton. It had only been a short time since the old man had agreed to it, and Davidson was eternally grateful for the older man's trust in his judgement. Lick had long been known for demanding loyalty and obedience from those he worked with and once asked workers on his orchard farm to plant fruit trees upside down to test their obedience. Anyone caught planting a tree right side up was summarily dismissed! After all, how dare they!

With his health fading and the future of his observatory secure, James Lick died on October 1, 1876, in his room at the Lick House. It would take twelve years for the observatory that bore his name to be finished. The 36-inch refractor in the main dome of the observatory was brought on line in 1888.

When finished the Great Lick Refractor (as it became known in later years) was the most powerful telescope in the world. The first major telescope and dome designed to be built out of steel, it was completed by the Union Iron Works of San Francisco and brought up the mountain's long winding dirt road by mule train. The giant lens sat at the end of a 55-foot tube mounted on a massive pier at the center of the 76-foot-diameter observing platform. At the beginning of an observing run an astronomer or assistant would climb the spiral staircase to the top of the pier and crank up an enormous stone, housed within the pier, to start the telescope clock movement. While the clock was engaged the telescope could easily be rotated using levers at the eyepiece to track whatever object the observer wished to follow. Another clever feature of the telescope was the dome floor, which was raised and lowered on a geared counterweight system to the height of the eyepiece for viewing. This feature meant that the astronomer didn't have to spend long hours at the top of a rickety ladder trying to chase the stars. When finished the rotating dome and slit weighed over 90 tons.

On January 8, 1887, the body of James Lick was exhumed from its plot at the Masonic cemetery and brought to Mount Hamilton, where it was laid to rest in a specially built chamber beneath the pier of the giant telescope. A plaque on the outside of the pier read: "Here lies the body of James Lick." The telescope and dome are still in use today.

****

George Ellery Hale peeled his steely-eyed gaze away from the enormous telescope standing before him long enough to check his pocket watch. The time was approaching 3 p.m., Central Standard Time, and he would have to leave the Manufacturer's Building shortly to join a meeting his father had asked him to attend. As quickly as it had been diverted, his attention returned to the work at hand, assembling the largest and most powerful telescope in the world. Hale knew there was little time to spare. The World's Fair Columbian Exposition of 1893 was less than a month away, and the native Chicagoan was eager to have the telescope ready for opening day.

Short and slender with short brown hair that was parted at the center, Hale had been frail and sickly as a boy. What he lacked in health and physical vigor he more than made up for in intelligence, curiosity and enthusiasm. A former student at the newly founded Massachusetts Institute of Technology Hale was fascinated by the field of astrophysics and was determined to make the study of stellar evolution his life's work. Working in his Kenwood Observatory, which he had built with his father's help on the grounds of his family home in Hyde Park, Illinois, Hale had invented the spectroheliograph. This instrument, which became the subject of his senior thesis, was designed to capture images of his favorite target, the Sun, in hopes of photographing the solar prominences and the corona without the aid of a solar eclipse. In Hale's eyes, the combined fields of astronomy and astrophysics were on the verge of remarkable advances, and he wanted to be at the center of developments within them. The giant refractor that was now being assembled before him was to be a major step in his quest to better understand of how the stars evolved. Hale hoped to show off his mighty new telescope for the millions of visitors who would soon be striding through the exhibits at the fair.

Destroyed by fire a little over two decades before, Chicago's downtown was now a shining example of the city's strength, ingenuity and resolve. Rising from the ashes and devastation of the Great Chicago Fire, the newly revitalized city had its place restored as the epicenter of American industrial growth. Stockyards at the center of downtown slaughtered millions of head of hogs and cattle per year. Due to innovations in refrigeration, meat could now be kept fresh while being shipped across the country on ice-packed railcars. Chicago was the hub for rail service in every direction. Steel, lumber, grain, livestock—everything that could be shipped over the rails from one end of the country to the other—usually went through the city's extensive railyards. Although it was not as robust as it had been just after the Great Fire, commercial construction was still strong. Some of the tallest buildings ever constructed rose above the horizon and could be seen for miles outside the city limits.

The ten-story Rand McNally Building, the world's first all-steel framed sky-scraper, was where George Ellery Hale was headed for his meeting. As he walked across the fairgrounds, Hale snuck a peek toward the lakefront at the enormous wheel George Ferris and Company were constructing. Designed by George Ferris, Jr., and built by the Bethlehem Steel Works in Pittsburgh, the great wheel would use massive steam engines installed beneath the fairgrounds to rotate 36 cars, each carrying 60 passengers around a 45-foot axle. The fair's organizing committee had commissioned the great wheel to rival the Eiffel Tower, which had been built for the World Fair in Paris several years before.

Hale made his way into the building, called for the elevator, waiting patiently for one of the cars to arrive. Although Hale elevators had been installed in high rises all over Chicago as well as the world since the middle of the 1870's, George Hale couldn't help reflecting on that fact every time he summoned one.

The headquarters of the fair's design and executive management teams were on the fifth floor of the building. In time the elevator car arrived, and Hale was gently spirited away up to his meeting. As he walked into the conference room Hale saw his father conferring with Charles Atwood, chief designer, and Daniel H. Burnham, the fair's lead architect and the man responsible for bringing the event to Chicago. It was at Burnham's insistence that fair organizers consider Chicago in the first place. Chicago was the epicenter for growth in post-Civil War America, and, in Burnham's mind, the only place suitable to hold an event commemorating the 400th anniversary of the country's discovery.

Daniel Hudson Burnham was a captivating pitchman who had established himself as a visionary designer in the 1880s. Tall, sturdy and handsome, Burnham could win over a room full of executives simply by stepping through the doorway. His partnership with John Wellborn Root led to several breakthroughs in architectural design and engineering. Two of the architectural firm's building designs, the Rookery Building (1886) and the Rand-McNally Building (1889), had established Burnham and Root as leaders in the world of architecture. While designing the buildings, Root began experimenting with the steel I-beams, a recent advance in building engineering. As a result of his tinkering the Rookery had a spacious light court that presented, for the first time, broad, open indoor spaces surrounded by glass without the need for columns to support the internal roof structure. The building also marked a turning point in building practices from the use of large masonry bearing walls to internal steel-framed construction. When it was finished the Rand-McNally Building became the world's first skyscraper built entirely using this latter method of construction. Capitalizing on Root's innovative genius and Burnham's inimitable prowess as a speaker and a front man, the architectural firm grew in stature, catapulting both men to the top of their field.

In 1889 France held a World's Fair in Paris, the Exposition Universelle, that mesmerized those who visited its grounds. The highlight of the fair was Alexandre Eiffel's iron tower. Standing almost 1000-feet tall, the tower was by far the tallest manmade structure ever built, and its presence eclipsed even the finest achievements of American designers and builders. Organizers and civic planners across the United States quickly began making plans to host a World's Fair that would top even the lofty heights the fair in Paris had achieved.

Several cities, including New York, St. Louis and Washington, D.C., made their cases for hosting the fair. Chicago, with a population now over a million people, the second most populace city in the country, was overlooked in the early going. Despite the efforts of planners and builders alike, Chicago was still seen as a second-class city, more butcher than baron and more beastly than beautiful. Burnham and Root thought differently, however. After all, they were the most respected architects in the country, and the city had shown its muscle as well as its creativity in rebuilding in the wake of the devastating fire two decades earlier. In earnest, Chicago's politicians began touting the city's virtues and, on February 14, 1890, their persistence was rewarded. Burnham and Root were in their offices on the top floor of the Rookery Building, at the corner of La Salle and Adams, when they heard the news. Swept up in the excitement and emotion of the moment, the two men immediately began making plans for the fair's location and overall complexion. Shortly after the announcement, a board of directors was named consisting of 45 men with Lyman Gage serving as president of the exposition. All of the men on the board were wealthy, and each had his own agenda. For eight months, progress was stalled while the board members argued over where to locate the fair within the city. Finally, as November neared, Burnham was made chief of construction. Burnham then appointed Root supervising architect and Fredrick Law Olmsted supervising landscape architect. Olmsted, who had designed both Central and Prospect Parks in New York City, was an admirer of Burnham's blunt but cordial style. A long-time friend and admirer, Burnham would later describe Olmsted's work in glowing praise: "An artist, he paints with lakes and wooded slopes; with lawns and banks and forest covered hills; with mountain sides and ocean views."

As a first course of action, Root created a collective of the best known architects to design the individual buildings that would house the thousands of attractions that would make the fair their home. Although well known in their own right, Root had the respect and admiration of every man he selected to the design team. With only thirty months remaining to locate, design and build the fairgrounds, everyone's complete cooperation and diligence was paramount to its success. By January 5, 1891, all ten architects chosen by Root, with Burnham's support, were on board. But months of travel, handwringing over appointments and decisions and restless nights wondering when they would actually be allowed to start the daunting project before them left Root exhausted. His health suddenly failed him, and on January 15, he succumbed to exhaustion and illness and died. The loss of his friend and collaborator for the past eighteen years crushed the normally high-spirited Burnham and left a void in the organization of the event. There was no replacing his friend, and Burnham knew he would be the one to take on the inordinately large task of managing both his and Root's workload.

In order to do so he chose Charles Atwood, a New York-based architect, to tend to the daily operations of Burnham's and Root's firm, leaving Burnham to manage the production of the fair. In addition to managing company contracts Atwood personally designed the Terminal Station and the Fine Arts Building on the fairgrounds. The latter today is the home of the Museum of Science and Industry.

Later, as a member of Burnham's staff, Atwood would design the Reliance Building (now the Burnham Hotel) and the Marshall Field and Company Building.

The death of John Root and the weather made the already tricky task of erecting and finishing the many state-of-the-art buildings in time for the fair all the more difficult. On June 13, 1892, the Manufacturer's and Liberal Arts Building, the very building in which George Hale was planning to display his telescope, collapsed during a storm. With the fair less than a year away, Burnham knew he had to commit even more manpower to the completion of the buildings and grounds. As a board member, William Hale sensed the urgency of the moment and enlisted his son to help expedite the remaining tasks in order to get the fair open by May 1, 1893.

Standing with William Hale and Burnham was Lyman Gage, the respected banker who had staked the elder Hale's fledgling elevator company after the Great Fire. In the midst of a recession, Gage had quit as president to attend to his banking duties but was still active on the board. As the group of men discussed the condition of the fairgrounds and buildings, the party was joined by Leander McCormick and Charles Tyson Yerkes. The two men had just been to the Manufacturer's Building so Yerkes could show off his giant new telescope to the elder McCormick. The inventor of a harvesting machine and co-founder of the McCormick Harvesting Machine Company (which would become International Harvester after its sale in 1902), McCormick had recently donated his own observatory to the University of Virginia. The 26-inch refractor had been the largest in the country until the Lick Observatory opened in 1888 with its 36-inch refractor. Three years retired from his harvesting company responsibilities, McCormick had vast land holdings in the area, while Yerkes owned most of the city's elevated railways. A known blackmailer who would think nothing of using a hefty bribe to get what he was after, Yerkes had decided only a year before to fund the younger Hale's dream. The energetic young astronomer had shrewdly convinced the aging railroad magnate that granting the funds for a new observatory and research facility, to be built in his name, would be a fitting and lasting tribute. Especially when it housed the world's largest telescope.

* * * * * * * * * * * * * * * * **

When the Columbian Exposition finally opened in May of that year, it did so to world acclaim, hosting over 27 million people. The White City, as it was called, featured numerous technological advances displayed in a series of beautifully designed white stone buildings with elegant waterways running through the center of the exhibit grounds, which covered more than 690 acres of the city's downtown.

The fairgrounds and buildings were lighted by incandescent light bulbs powered by alternating current, the first time these had been in large-scale use. Other features of the fair included the Eastman Company's Kodak 4 camera, which looked like a brown box with a lens in one side, the first night football game and a long moving sidewalk with benches that escorted seated fairgoers to the foot of a long pier on the lake front. Scott Joplin's new ragtime-style music was introduced to the masses. Beautiful bridged waterways wound their way through the fairgrounds, and guests could ride aboard a replica of a Viking ship to tour the grounds. Elsewhere, an exhibit by the Krupp Company of Berlin displayed a new type of cannon weighing 122 metric tons and capable of firing an explosive projectile at a target miles away with deadly precision.

The fair had a lasting impact on the American psyche and helped to usher in the City Beautiful movement that sought to elevate the country's cities to the levels of the great cities of Europe. The success of the event and its impact made Burnham the country's best known architect. Having stepped out of the shadow of his late talented partner, John Root, in the years to come Burnham would create city plans for Cleveland and San Francisco and help to design the mall in Washington, D.C., as well as Chicago's Miracle Mile. Two other Chicago monuments, Soldier Field and the Field Museum, were designed by Burnham. Another of his designs, the Fuller Building in New York, later became known as the Flatiron Building, so named for its resemblance to a common household appliance.

Inside the Manufacturer's Building, the 25-year-old observatory director, George Ellery Hale, was presenting another marvel of science, the 40-inch Yerkes refracting telescope, to the world. The instrument was proudly displayed mounted on its 50-ton pier over four stories above the floor in a prominent corner of the exhibit area. The polar and declination axes and drive gear, weighing an additional 25 tons, were mounted on top of the pier beside the telescope's tube. The front of the 62-foot tube housed a 40-inch lens crafted by Alvin Clark, and the instrument was guided using electricity, the first large telescope ever to be driven that way. Its 36-inch counterpart at the Lick Observatory, completed just years before, used a clock driven by a large stone that had to be wound up every few hours.

Once installed at Yerkes Observatory in Wisconsin, in 1897, the great refractor would be centered inside a 90-foot-diameter dome. Just as the Lick refractor had been designed to do, the entire 75-foot-diameter floor surrounding the Yerkes refractor, which weighed almost 38 tons, was raised and lowered on four motorized cables to get the observer into position beneath the eyepiece. The telescope mounting, dome and floor was designed by Warner and Swasey. So heavy was the floor, that in the early morning of May 29, 1897, it collapsed into a pile of rubble at the bottom of the dome. This was a situation where working the late shift had its advantages. Even though astronomers work in the evening, they had all retreated after a long night's work and were safely in their beds when the dome floor came crashing down. Ferdinand Ellerman, a stalwart stellar and solar observer, fix-it man, and staff photographer at Yerkes and later Mount Wilson Observatory, was the last to leave the room before the collapse. He reported hearing something snap like a piece of wood as he tried to raise the floor at the end of his observing run. Ellerman said that he had switched off motors and checked for damage but could find none. Hours later the floor lay in a heap at the bottom of the dome. Hale and the others were summoned, and Warner and Swasey were called to the observatory to assess the damage to the telescope and pier. The telescope was fine and the pier had received only minimal damage, so the observing floor was rebuilt under an improved design and put to good use for years to come.

For young George Hale, the design and creation of this great new instrument was an incredible victory. As a boy he had unleashed his boundless curiosity at the nature of the cosmos and, aided by his father's endless support, had continued to improve not only his knowledge and technical skills but put the equipment he was using to good use. Vexed over the problem of studying the solar prominences, which had been impossible without the aid of a solar eclipse, he had devised and created his spectroheliograph to allow him the ability to study his favorite stellar object, the Sun, on a daily basis. Now, at a time when technology was allowing for ever-greater instruments to be built, Hale had seized the initiative, creating the largest working telescope in the world. On top of all that, he had proven himself a worthy negotiator, having talked the cantankerous millionaire rail magnate, Charles Yerkes, into funding the creation of a state-of-the-art observatory near Lake Geneva under the auspices of the University of Chicago.

In Hale's mind this was just the beginning of things to come. Already his chief telescope designer, George Ritchey, was building a 24-inch reflecting telescope, which used a 2-foot primary mirror at its base and a secondary mirror mounted near the end of its tube to reflect light in an ever-decreasing funnel through an eyepiece where the observer could view his target. The new telescope design increased focal range without the need to increase the size of the telescope tube significantly, and Hale suspected these telescopes would come to dominate the field of astronomy in the future. In 1894, while discussing the problem of creating even greater mirrors for use in larger and more powerful telescopes, William Hale agreed to commission a 5-foot glass disc that could serve as the primary mirror in a giant new telescope. In 1896 the St. Gobain Glass Works in 1France created and shipped to the lab at the Yerkes Observatory the 8-inch thick, 1-ton glass disc. Hale had the blank he needed to once again design and build a telescope that would eclipse even the great 40-inch reflector. He still needed the location and the funding to build it. He dreamed of the day the 60-inch would see first light, a dream that would be twelve years in the making.

* * * * * * * * * * * * * * * * * * **

On New Year's Eve, 1901, Andrew Carnegie received a letter from President Theodore Roosevelt at his home in New York City. The new president was writing in praise of the recently conceived Carnegie Institution of Washington with his pledge to join as a member of its prestigious board of trustees. As president, Roosevelt would become a board member by virtue of his office according to the Carnegie Trust. Other board members included Secretary of State, John Hay, who had served as assistant to Abraham Lincoln during the Civil War, Secretary of the Treasury, Lyman Gage and former president, Grover Cleveland, who had been the first and only Democrat elected to the nation's highest office since the end of the war between the states.

The usually optimistic Carnegie had found himself depressed and cynical at the thought of Roosevelt, who he regarded as "dangerous," ascending to the presidency. Carnegie's views put the republic in Republican. He had opposed the decision of Roosevelt's predecessor, William McKinley, to assume control of Cuba (occupied only until 1902), Puerto Rico, Guam and the Philippines after U.S. armed forces had routed Spain in the 1898 Spanish-American War.

In spite of his differences with McKinley over U.S. imperialism in the region after the war, Carnegie had endorsed his re-election as president in 1900 over his opponent William Jennings Bryan, whom he saw as a menace to the homeland.

McKinley won re-election easily over Bryan and was sworn in with Roosevelt as his vice-president. Just six months into his second term, McKinley was assassinated by a young insular Polish copycat killer named Leon Czolgosz. The 28-year old steel worker from Ohio had followed the president to Buffalo, New York, the site of the Pan-American Exposition, where he shot McKinley twice in the chest with a revolver at close range. Czolgosz said he was acting on behalf of the common people, who he felt were being exploited by the rich, and he blamed the government for the injustice. Carnegie was deeply saddened by the news. "President McKinley gone—Isn't it dreadful," he wrote to a friend. For his trouble, Czolgosz was executed by electrocution 45 days after McKinley's death on September 14, 1901.

Roosevelt was sworn in that day, becoming the 26th U.S. president. Carnegie confessed he was, "not very confident about Roosevelt's wisdom," but thought that "power may sober him." In spite of his misgivings, Carnegie sent the young president his plans for his new philanthropic institution on Thanksgiving Day with a note expressing his joy in having the opportunity "to prove, at least in some degree my gratitude to, and love for, the Republic to which I owe so much."

Carnegie's intention was to give the $10 million (over $275 million today) to the government in keeping with his desire to redistribute his fortune before he died rather than pass it on to his family. He was convinced the time-honored practice of patrimony was irresponsible and decried the practice in his article, "The Gospel of Wealth," published in *The North American Review* in June of 1889.

Not wanting to invite a public outcry for taking money from the sale of a private corporation, especially the controversial U.S. Steel, Roosevelt declined the steel magnate's offer. In response to the government's decision, Carnegie organized the institution that would bear his name under a trust.

Although sizable by any standard of that or any other period in history, Carnegie's endowment was but a small fraction of his worth.

A few months earlier Carnegie had accepted an offer from a consortium of interested businessmen, led by J.P. Morgan, to sell his Carnegie Steel Works for $480 million (over $13 billion today). The deal, which led to the creation of the U. S. Steel Works, made Carnegie one of the richest men in the world. But questions about his motives and practices in attaining his fortune still lingered in the public sphere, and now Carnegie was more determined than ever to dispel public distrust.

The Scottish-born Carnegie had moved to the United States from Dunfermline in the 1840s. He was twenty-five years old and the Western Superintendent of the Pennsylvania Railroad Company when war broke out between the Union and the Confederacy in 1861. Short and lean with youthful looks, a chin curtain beard and short-cropped, wavy brown hair that was parted to one side, Carnegie bore a look of steel-eyed confidence. He was an eloquent and unabashed skeptic of southern politics and lent his full support to any effort to end the oppression of African Americans in the South.

Carnegie profited greatly from the Lincoln government during the Civil War, supplying steel for Union weaponry and helping to transport and arm Northern troops. In the fall of 1863 Carnegie jotted down his earnings on office letterhead. His interests in oil and companies supporting the railroad industry would earn him $47,860.67 (about $1.3 million today) that year.

In 1865, as the war was coming to a close, Carnegie formed the Union Iron Mills through a partnership with a nearby steel manufacturer, appointing himself as president. While Reconstruction was getting under way in an effort to restore peace in the Union after the war, steel was replacing wood in the design and creation of new railway bridges and buildings. The completion of the world's first cross-country railroad led to greater western expansion as new cities grew out of western soil from Iowa to California. Seizing on the opportunity, Carnegie worked to refine his steel-making process.

In 1876, as the world reeled under three years of economic depression, events were unfolding that would reverberate on the national and world stage for decades to come. On March 10, Alexander Graham Bell spoke into a primitive telephone receiver and said to his assistant, Thomas Watson, who was listening on the other end, "Mr. Watson—come here—I want to see you." And thus began the era of telephone dating. The National League of Base Ball Clubs was established in April with eight teams, including two teams that are still in operation today: the Chicago Cubs (then known as the Chicago White Stockings) and the Atlanta Braves (then called the Boston Red Stockings). After a pyrrhic victory over U.S. forces at the Battle of the Little Bighorn in June, Sitting Bull and his fellow tribesmen and women were hightailing it over the Canadian border with U.S. reinforcements in hot pursuit. After battling starvation and the bitter Canadian winter, Sitting Bull would lead his people back to U.S. soil in disgrace. The last battle had been won, but the war was over and all was lost. After losing their first popular election in twenty years, Republicans were negotiating a settlement with Democrats in Congress to maintain their grip on the presidency. The Compromise of 1877 would give their candidate, Rutherford B. Hayes, the 20 electoral votes disputed in the election and the presidency, while ceding power in the south to southern Democrats, restoring white supremacy in the region and effectively ending the era of Reconstruction.

Meanwhile, Andrew Carnegie was, not so quietly, building the greatest industrial empire of the twentieth century. He purchased controlling interest in a competing steel works that used a state-of-the-art steel processor known as a Bessemer converter, further refining his steel processing plants. The newly acquired plant also offered extensive railways that he extended to combine and connect his various steelyards with his suppliers. Furthering the vertical integration of his operation, Carnegie bought coke and other mineral interests until he held control of the entire supply chain. This allowed him to capitalize fully on the burgeoning steel industry's expansion. In time his dogged determination, ruthless business management and skillful navigation of the fluctuating economic landscape would lead Carnegie to create the world's largest company.

Carnegie's innate ability to win people's favor enabled him to reap a fortune on the backs of men working twelve hours a day, seven days a week for meager pay. When union negotiations failed, as happened late in 1874 with the Amalgamated Association of Iron and Steel Workers, Carnegie used his considerable wealth to wait out and eventually defeat them, ending the strike and sending workers back to work at the mills. In time rifts between labor and management sometimes erupted, resulting in one of the stormiest and most violent union uprisings in the country's history.

One incident, involving Homestead Steel Works and the Amalgamated union in the summer of 1892 tarnished Carnegie's reputation and signaled the decline of organized labor in the United States until the 1930s. Amalgamated union workers, disgruntled at the working conditions and hours at the Homestead plant, went on strike, demanding contract renegotiations. Carnegie and his business partner, James Frick, were against the union's heavy handed tactics and refused to budge. Disputes between the two sides had been ongoing since 1889, but to date Carnegie had put down any attempts using his deep pockets to wait out union workers and a large contingent of Pinkerton detectives to discourage any organized attempt to cause physical harm to the plant and grounds. In Carnegie's opinion, the union was already making it difficult for the firm to control costs in production, and he sought to break the union by any means possible. Attempts were made to bring in skilled labor to operate the mill from nearby towns, but striking union mill workers routinely located the scab laborers and escorted them, often unceremoniously, back out of town. Townspeople who witnessed the bullying of the men who were coming to town in the hopes of finding work began to dislike the union's tactics, even as they supported their cause.

Finally, while Carnegie was away at his Skibo residence in Scotland, a band of 300 Pinkerton detectives was brought in to drive off the striking union workers and allow strikebreakers access to the grounds. This time the union had had enough and vowed to fight. Early on the morning of July 6 the Pinkerton detectives and strikers squared off in a series of skirmishes and gunfights that lasted twelve hours. Seven strikers and three of the detectives were killed, and dozens more were wounded on both sides. In the end, the Pinkerton detectives surrendered to some 3000 striking workers. The Pinkerton men were beaten as they were marched to the Opera House in the center of town that was being used as a temporary jail to house them.

The Pennsylvania governor dispatched 6000 of the state's militia to restore order to the area. Within minutes of its arrival the militia established martial law, bringing an end to one of the bloodiest labor rights battles in the country's history.

The treatment of the defenseless Pinkerton detectives turned public sentiment against the union strikers. This and the cost of legal battles left the union financially and morally bankrupt. The Amalgamated soon disbanded, and Carnegie Steel as well as many other plants in the area and around the country were restored to non-union workers, and unions would not gain a foothold in the American workforce again for forty years.

Although he had won the battle for control of his steel works, Carnegie had lost another battle, one more precious to him than any other. Union blood spilled during the battle for the Homestead Steel Works had tarnished Carnegie's reputation as a man of patience and magnanimity. Although Homestead grew into the world's largest and most technically advanced steel manufacturer, Carnegie would spend the rest of his life repairing the damage the Homestead strike had caused to his reputation.

Ten years later, while in his sixties, the inherently optimistic Carnegie was embarking on a new career in philanthropy. His take from the sale of his steel works was roughly $320 million, and Carnegie set out to donate all but a small portion of his fortune before his death. This proved difficult even for the ambitious philanthropist. At the time of his death in 1919, Carnegie would be worth roughly $475 million (now over $13 billion), making him one of the richest men in history. He would manage to redistribute all but about $30 million of his wealth back to the public sector through various charitable trusts and foundations. Carnegie's giving always put an emphasis on learning and education, and he created some of the finest learning facilities in the country. Among these was the Carnegie Institution of Washington, which Carnegie founded in January of 1902, with an endowment of $10 million. In his trust, which was written in the pages of the inaugural yearbook of the Carnegie Institution of Washington, Carnegie spells out the mission of his nascent foundation. Among the aims Carnegie listed in his charter was: "To discover the exceptional man in every department of study whenever and wherever found, inside or outside of schools, and enable him to make the work for which he seems specially designed his life work."

That year 10-year-old Milton Humason boarded a train car on the transcontinental railroad with his family, bound for Los Angeles, California.

Part I
An Ordinary Life (1891–1917)

Obscurity and a competence—that is the life that is best worth living.

Mark Twain

Chapter 1
Early Childhood in Minnesota

Abstract Born amid the extraordinary circumstances shaping the socio-political landscape in the United States during the last decade of the nineteenth century, Milton Humason grows up in the bustling town of Winona, Minnesota, one of three children born to William and Laura Humason. War, the assassination of President William McKinley and family tragedy help to alter the course of the family's history by the end of 1901.

Just after dawn on December 29, 1890, a mounted force of 500 soldiers of the U.S. 7th Cavalry, led by Colonel James Forsyth, rode into an encampment of Indians led by the Miniconjou chief, Big Foot, at Wounded Knee Creek. The army had intercepted the small band of Miniconjou five days earlier 30 miles east of the Pine Ridge reservation, where they had hoped to rendezvous with Short Bull, Kicking Bear and their followers at a remote outcropping of steep rock called the Stronghold. The tribes were moving in protest of the ongoing mistreatment of their people at the reservations.

Two weeks earlier the Lakota holy man, Sitting Bull, and seven of his warriors had been gunned down by Indian police in his home at the Standing Rock reservation. Six of the police officers were also killed in the violence, and the incident left the nearby native population in a fury. The slain tribal leader was well respected and was among those who helped band their tribes together in recent years to ward off the incursion of federal troops and white settlers on their lands. One of the heroes of the decisive Indian victory over Lt. Colonel George Custer and the 7th Cavalry at the Battle of Little Bighorn in 1876, Sitting Bull and his followers had been pursued by U.S. forces and fled over the Canadian border, where they remained until 1881. At that time, abandoned by the Canadian government and suffering from the harsh Canadian winters, Sitting Bull had brought his people back to the reservation on U.S. soil and surrendered. Later he had found fame and some fortune touring with Buffalo Bill Cody's Wild West Show in 1885 but quit after months on tour with the show, returning to his home at the Standing Rock reservation, where he redistributed most of the money he'd earned among his people.

In the aftermath of Sitting Bull's death Big Foot, Kicking Bear and Short Bull defiantly moved their tribes off their reservations. Along the journey Big Foot had

© Springer Science+Business Media New York 2016
R.L. Voller, *The Muleskinner and the Stars*,
Springer Biographies, DOI 10.1007/978-1-4939-2880-4_1

contracted pneumonia and was relegated to riding in the back of a wagon. The party slowed and was intercepted. They never made it to the Stronghold.

Forsyth's mounted force was backed by four Hotchkiss cannons positioned on high ground surrounding the camp of 350 Indians, two-thirds of whom were women and children. The air was cold and crisp and the sky was a steel gray as a heavy snow threatened to sweep in from the north. As so often happened in cases involving Indians and armed U.S. forces, who fired first was never really known, but the outcome had become all too familiar. Forsyth demanded the Indians surrender their weapons and return to the reservation peacefully. All but one, a deaf man named Black Coyote, did so quietly. When an officer tried to take his gun, Black Coyote resisted and the gun went off. In an instant the tense but peaceful confrontation erupted in hand to hand combat. Outnumbered three to one by their would-be captors, the Indians began to run for cover. The Hotchkiss guns were unleashed on the fleeing Indians, strafing them with a hail of cannon and gunfire. Less than an hour later, 200 members of the tribes, mainly women and children and the elderly, lay dead or wounded in the cold winter air. Big Foot had been among the first to fall. Army casualties listed as 25 killed and 39 wounded. Public sentiment for the plight of the Indians swelled in light of the massacre. Forsyth was later tried for murder in the incident but was exonerated.

In the days and months that followed, as the echo of the big guns and the emotive cries of innocents fell silent on the northern plains, winter turned to spring and spring into summer. On August 19, 1891, 600 miles east in a town called Dodge Center, Minnesota, Milton La Salle Humason was born.

In perhaps a strange coincidence of fate, on that same day the 67-year-old spectroscopist, Sir William Huggins, opened the proceedings of the conference of the British Association for the Advancement of Science in Cardiff, Wales. James Keeler, who had resigned as director of the Lick Observatory in June, referred to Huggins in a note to his successor, William Campbell, as "the founder of the science of astronomical spectroscopy." Of the early spectroscopists, Huggins had done more than perhaps anyone else to advance the emerging field of astrophysics. He had been the first to observe the emission lines in the spectra of nebulae, to use the Doppler principle to determine stellar motion along the line of sight and identify the ultraviolet lines in the spectra of hydrogen on film. All of these practices, Milton Humason would one day put to good use in helping to advance the field even farther.

In his speech to the B.A.A.S. Huggins, who, along with his wife Margaret was far from finished in contributing to the advancement of the field, discussed the existing and emerging developments of the moment, with a nod to the spirited group of young men who would continue the research on stellar evolution into the future. "Happy is the lot of those who are still on the eastern side of life's meridian," the newly elected president said in closing. His comments were directed largely at young George Hale, who was seated in the audience that night and was set to give a presentation of his attempts to photograph solar prominences the following day. The two had met a week earlier at the Huggins' observatory in the Tulse Hill district of London, and Hale had delighted his older counterpart with his energy and enthusiasm, both for the field of astrophysics and for Huggins' contributions to it.

The massacre at Wounded Knee and the B.A.A.S. conference in Wales illustrate the extraordinarily dichotomous times in the United States into which Milton Humason was born—a new world brimming with opportunity and industrialization straddling an antiquated yet prideful world, banished ignominiously to the margins of society. In his eight decades Humason would live through the greatest period of modernization in history, one that began with the advent of the telephone and the lightbulb and ended with men walking on the Moon. And although he always greeted the world he lived in with a handshake, a smile, and a story or a joke, his love for that antiquated world never waned (Fig. 1.1).

Humason's ordinary birth came during an extraordinary period in the growth of the United States as a nation. With the nomadic tribes of the northern and southern plains now contained on reservations, European emigrants were freer than ever to push west in the quest for land and opportunity. Western expansion, the discovery of oil and gold, and advances in building technologies were creating unprecedented wealth. The steam engine had become the workhorse of the surging Industrial

Fig. 1.1 A portrait of Milton Humason, circa 1894

Revolution while telephones, electricity and the lightbulb were enhancing both work and living environments. Victorian Era charm and grace were being replaced by the grimy advance of industry both in Europe and at home in America.

Milton Humason was the oldest son of William Grant Humason and Laura Petterson Humason. The Humason family dated back to before the Revolutionary War. Several of William Humason's descendants served their young nation during the Revolution and the War of 1812. Through the years the family moved steadily westward with the country's expansion. William's father, Lewis Abel Humason, was born in Ohio in 1837. In 1862, he married his cousin, Ellen Amelia Humason, in Rochester, Minnesota, and the couple settled down while Lewis took work as a miller during the early part of the Civil War. In June of 1864 infantry pay went from $13 per month to $16. Seizing on the opportunity Lewis left his pregnant wife to enter the war in the hopes of making a little extra money while serving his country, as his ancestors had done before him. William was born shortly after the war ended and, in 1873, as a world economic crisis began to grip the country, Lewis decided to move the family west to San Francisco, where jobs would be more plentiful.

Laura Petterson Humason was a second- generation American whose paternal grandparents had come over from England shortly after getting married sometime in the 1830s. Laura's father, George Howard Petterson, had been born in New York and her mother, Mary Catherine Shirkey, in Virginia. George and Mary had five children. Three of them, Sarah, Charlotte and George, Jr., were born in New York. Young George's birth coincided with the start of the Civil War. There is some evidence that exists suggesting that George Petterson senior entered the war. In any case, the Petersons were living with three children as the war broke out between the Union and Confederate armies, and shortly after the war ended, they were living in San Francisco with two more daughters, Laura, born 1868, and Alice, born 1870. As there were seven years separating them from their older siblings, Laura and Alice naturally grew up together and learned to take care of each other. This bond would show its true strength later through misfortune and family tragedy.

William and Laura were 21 years old when they married in San Francisco in 1889. Soon after their marriage the young couple moved east to Dodge Center near William's childhood home of Stewartville, Minnesota. Lewis Humason, William's father, ailing from injuries suffered during the War Between the States, had decided to move back to his hometown to live out his days with the rest of his close relatives. William asked Laura if she would move there with him so that he could tend to his father, and Laura consented.

A year later, Milton Humason was born. He got his odd middle name in honor of the man who had saved his father's life. Soon after their move to Stewartville, William fell into an abandoned well near the family's home, injuring himself, and was rescued by a man named La Salle. William and Laura named their first born after the man in recognition of the debt they felt was owed for his kind and unselfish act. A little over two years later Milton's brother Lewis Howard was born on September 7, 1893.

For nearly two decades much of the country and the world had been mired in a long economic decline (sometimes called the Long Depression due to how long it lasted, 1873–1896). It was referred to as the Great Depression until the shorter but more severe depression of the 1930s. Fears of a bubble had caused stock exchanges to collapse after a panic in Vienna set off a run on its banks in 1873. The ripple effect from the event made its way across Europe and finally gripped the United States on September 18, a day that was known as Black Thursday. The crisis in America was set off by a railroad industry bubble that finally burst after the banking house of Jay Cooke and Company failed to acquire a $100,000,000 bond for the Northern Pacific Railroad.

The railroad had sought to expand onto 40 million acres of public land it had acquired, but when the bond proved to be unsalable the bank collapsed. Within days several other major banks also collapsed, causing the New York Stock Exchange to close for ten days. Fears of a manufacturing bubble and corruption in the railroad industry along with the government's attempt to get back to the gold standard fanned the flames of inflation and drove unemployment to nearly 14 %. Adherence to the gold standard helped the U.S. economy begin its recovery by 1879, but the return to increased productivity was felt more in the big cities than in rural areas. In these areas recovery was slow and often painful and lasted well into the 1890s. As a result Dodge Center, like many small towns across the country, experienced population decline as people fled the small community of 600 residents for large cities such as Winona and St. Paul. William and Laura, by no means immune to the fallout from the Depression, moved their young family to Winona in 1892.

Winona's economy, based on the grain and lumber industries, had been thriving until about 1890, when it went into a slow but steady decline. Still, along with its larger neighbor to the north, it was the center of industrial might in the state. Set against the shore of the Mississippi River Winona was still one of the largest wheat shipping ports in the country. The river teemed with barges picking up grain shipments to run downriver. Massive timber rafts—long chains of trees felled in forests upstream and tied together into large floating rafts—were driven downriver into port, where mills waited to convert them to building materials. Traincars were loaded with grain, lumber, textiles and other goods and supplies for distribution to towns and cities across the region. Factories along the river buzzed daily with activity breathing life into the local economy. The Humasons were well entrenched in the local economy, so educated and capable William had no trouble finding work. Shortly after the couple landed in their new home, he took a job as a bookkeeper at the Minnesota Elevator Company at the Grain and Lumber Exchange Building. The building was within easy walking distance from the family's rented two-story home on Center Street.

In the decades since the Civil War the United States struggled to find its footing, as public pressure slowly mounted against the government to create jobs, heal the inequities in the labor market and end the oppression of African Americans in the South. The passivity of one leader after another combined with the expansion of the country, and the government's own internal battle for power left the public feeling abandoned and rebellious. Labor strikes and riots became more and more commonplace and occasionally boiled over into extreme violence. Political unrest existed both between and within the parties. The much disputed election of 1876, in which Democrats in Congress effectively sold electoral votes to the Republicans, gave Republican candidate Rutherford B. Hayes the presidency in exchange for the restoration of Democratic power in the South. The Compromise of 1877 had led to the end of Reconstruction and a return to oppressive prewar policies in the region. As a result black voters in the South were often denied access to voting booths during election cycles. In 1884 Democrats used this and other tactics to get Grover Cleveland elected president. An effective orator but an ineffective leader, Cleveland managed to do two things that had not been accomplished by a sitting president before or since: he was married in the presidential mansion to the young daughter of his former law partner, and he would become the only president elected to two non-consecutive terms.

By the end of the 1880s the most contentious Congress in U.S. political history was developing as power was becoming evenly distributed between the two sides. After long ruling the political landscape Republicans were beginning to take on the complexion of a party that was falling behind. The party's growing lack of perception of the desires of its constituents was leading to infighting and greater party despair. Meanwhile, faced with a golden opportunity to capitalize on the failures of its enemy across the aisle, Democrats in Congress had no idea how to capitalize on it. Meaningful and uplifting public events, such as the dedication of the Stature of Liberty in 1886, were undermined and overshadowed by events such as the Haymarket riots between striking workers at the McCormick Harvester Company and police, in Chicago, that killed or wounded scores of policemen and protesters. The dedication of Yosemite as a national park in 1890 was offset by the massacre of innocent Native Americans at Wounded Knee Creek later that year. The opening of the Columbian Exposition in 1893 was offset by a panic in the markets that sank the country into depression. These events are easy to conflate too strongly when viewed through the long lens of history, but their effects, however weakly connected, were felt strongly across the country and the world (Fig. 1.2).

However, for young Milton Humason and his little brother Lewis, growing up in the sprawling town of Winona on the shore of the Mississippi River, was ideal. By any kid's standards, Winona held all the promise and wonder the world had to offer. The city was intersected by several train lines and a streetcar that ran passengers through the center of town along 3rd and 5th Streets. The Chicago and Milwaukee Line ran along the eastern edge of the city near Lake Winona, dropping passengers at the depot on Bellevue Avenue. A line on the far side of the lake intersected the bridge spanning the lake from Huff Street in town. The bridge was the most direct route to Woodlawn Cemetery, nestled into the base of the flat hills on the far side of the lake. As was often the case in river cities, freight was shipped into the rail yards along the riverfront. The swing bridge across the Mississippi River made it possible to ship materials in and out of the city directly into Wisconsin. The bridge rotated on its axis to allow boat traffic up and down the river. A turn-around built into the rail yards in town made it easy to switch directions moving in and out of the city limits. Lumber mills and grain silos lined the shore, their small communities of laborers and families housed within the limits of the factories. The vast park near

Fig. 1.2 Two-year old Milton Humason stands in the yard of the family home on Center Street in Winona, Minnesota, in 1893. His infant brother, Lewis, sits on his mother's lap on the porch

Lake Winona played host to fairs, political campaigns, parades, baseball games and horseshoe tournaments. Paddlewheel boats were docked at the pier along the river for those who sought more adult entertainment. All night the gleaming lights and live music could be heard on the riverfront as gamblers tried their hands at the tables or availed themselves of the company of the ladies on board.

From an early age, Milton was dazzled by the clattering romp and rumpus the paddleboats abruptly infused into the otherwise bucolic surroundings of the city. Lean and of average height, with a thick head of wavy brown hair, Milton's bright blue eyes twinkled with mischief from an early age. At night he would sometimes lead Lewis down to the riverfront and sneak onto the boats, posing as young entrepreneurs, to get a glimpse of the nightlife inside the billiards and poker rooms. Charmed by his wit and daring, local gamblers often sat Milton down with them so he could view the games first hand. Thus began a lifelong affinity for the glitter and swagger of the life of a riverboat gambler.

Milton's other lifelong passion started around this time as well. As a boy, his father taught him and his brother the art of fly fishing and hunting. As with most things, Milton learned quickly, showing an aptitude and thoroughness of detail that would come to define him later in life. He learned to tie flies and began to excel at coaxing fish to bite at them at an early age. Fishing the nearby rivers and streams became a favorite pastime for the boys in those early years leading up to the turn of the century.

As a student, Milton applied himself diligently to his study work, although he enjoyed reading and arithmetic over writing. His favorite books, *The Jungle Book,*

Treasure Island and *A Christmas Carol* appealed to his adventurous heart and became lifelong companions handed down to his grandchildren. These books informed Milton's natural ability for storytelling, a device he would use throughout his life to defuse tense situations and bring smiles to the faces of those who knew him.

In 1897, Laura Humason received a letter from her little sister, Alice, informing her of her intentions to marry a Los Angeles, California, banker named Henry Clayton Witmer. The influential banking and real estate business owner had met Alice on a trip to San Francisco a couple years before, and the two began a long courtship shortly thereafter. Alice, who was reticent at first, had been confiding in her sister since she met Witmer, and lately it had seemed to Laura as if her she would finally consent to the persistent millionaire's wishes. Eager to respond, Laura quickly sent a telegram to her sister conveying her hearty congratulations and assurances that all would be well for her beloved sister and her husband to be, who was fourteen years her senior.

* * * * * * * * * * * * * * * * * **

The election of 1896 had brought the Republican nominee, William McKinley, to the presidency. McKinley inherited a country in the throes of the worst depression in its young history. His predecessor, Grover Cleveland, had proved again in his second term that inaction was poor policy, and McKinley was determined to change the fortunes of his countrymen for the better. The most progressive president in almost forty years, McKinley would move the country out of depression and land it squarely on the world stage. He was the first president to have his inaugural address put on film and set up the press room at the Oval Office where meetings between him and his cabinet members and the press became commonplace. He provided a telegraph for reporters to easily transfer the information gleaned from these meetings to their press corps. McKinley worked early in his term to increase U.S. business concerns abroad in the hopes of moving the country from a continental to a global mindset. His eye was on the annexation of Hawaii, and soon events unfolded that would enable McKinley to do just that.

In January of 1898, Alice Petterson became Mrs. Henry Clayton Witmer in a small marriage ceremony in Los Angeles. Shortly after, Laura Humason received a letter with a beautiful photograph of her sister in her wedding dress. After the recent and sudden deaths of both her husband's younger brother and brother in law, the couple's union was celebrated by the family as a sign of new beginnings.

As the Humason family's eyes were on the newlywed couple, the eyes of the nation were turned toward the Spanish territories of Cuba, Puerto Rico, Guam and the Philippines. After thirty years of unrest Cuba had once again erupted in rebellion. Sensing Spain's inability to put down the rebellion in its remote territory, McKinley dispatched the warship, *Maine,* to Havana on a goodwill mission to provide protection for U.S. citizens. On February 15, 1898, an explosion on the *Maine* sank her in Havana Harbor, killing more than 250 of the sailors on board. Although the cause of the explosion was unknown yellow journalists such as

William Randolph Hearst published articles intended to increase public anger at the Spanish government. Proceeding cautiously, McKinley telegrammed Admiral George Dewey, commander of the Pacific fleet, to prepare for an attack on the Philippines. To buy time, McKinley ordered an investigation into the cause of the sinking of the *Maine,* while the press fomented public support for war. The results of the investigation revealed on March 28 that an outside force, probably a mine from a submarine, had sunk the new U.S. warship. (A follow-up investigation almost eighty years later revealed that an explosion in the engine room probably caused the ship to sink.)

McKinley had the ammunition he needed. He sent a letter to Spain demanding an armistice until October 1, the immediate closure of Spanish concentration camps and the start of negotiations toward bringing about a permanent peace arbitrated by McKinley and the American government. Spain could not acquiesce to the last point for fear it would cause an uprising at home. After some internal negotiation surrounding the president's refusal to recognize a possibly radical revolutionary government in Cuba, McKinley asked for and was granted a declaration of war against Spain, issuing a call for 125,000 volunteers on April 25.

William G. Humason was among the first to enlist. The 29-year-old bookkeeper and father of two had been following the reports of the treatment of citizens in the Spanish territories and had determined to help do something about it. The descendant of a long line of soldiers, William heard the call of duty very clearly. He was mustered in with Company C of the 12th Minnesota Volunteers, led by Capt. Lincoln Gallien on May 6 and soon arrived at Camp Ramsey in St. Paul. The camp was set up at the Minnesota State Fairgrounds, and the 12th was housed at the horse and cattle barns during its stay. What William didn't know at the time was that Laura was pregnant with their third child, a fact that might have altered his decision making. Once at the camp, the regiment began maneuvers to ready itself for battle. The 12th boarded a train on May 16 bound for Camp George H. Thomas, Chicamauga National Park in Georgia, under the command of Gen. John A. Brooks, and arrived on May 19. The regiment began drills under Lieut. Col. Frank B. McCoy on its first day in camp.

The conditions at the camp were brutal. Insufficient food and water combined with bad weather and living conditions led to an outbreak of typhoid fever, which William contracted in late August. At the time no effective vaccine for the disease (which causes fever, severe diarrhea and intestinal hemorrhaging) existed and it was often fatal. William was given a medical furlough on September 2nd and sent to St. Luke's Hospital in Chicago. With the family's future in jeopardy, he lingered there for months struggling to overcome the disease.

The "splendid little war," as Secretary of State John Hay called it, began when the U.S. navy, commanded by Admiral Dewey, cornered and sank the Spanish fleet at Manila. It ended 3 months later. It was the easiest and most profitable war in U.S. history and led to the occupation of Cuba (temporarily), Guam, Puerto Rico and the Philippines. Fewer than 2500 Americans lost their lives, but only 385 of those men died in battle. The rest were taken largely by disease. For William Humason, victory in the war was overshadowed by his own battle to survive typhoid and regain his health, one that would continue for years to come.

****

With William still convalescing in the hospital in Chicago, Laura gave birth to the couple's third child, Virginia, on January 8, 1899. Early in February the middle part of the country was gripped by a blizzard and one of the worst cold spells in recorded history. From February 7–12, the cold front swept in, blanketing the country in record cold all the way to Texas. The mean temperature in Winona during the six-day period was −20. It was so cold that ice formed on the Mississippi River and flowed downriver until it reached the Gulf of Mexico.

Finally, the brutal winter turned to spring, and, in April William was cleared to leave the hospital. He boarded a train at Union Station in Chicago bound for Winona where Laura, with Milton, Lewis and 4-month-old Virginia in tow, met him at the train depot. The year away and his long illness had taken a toll on William. His weakened appearance was startling, and he was susceptible to cough, headaches and stomach flu, and the spring rains offered him no reprieve from his ailments.

On June 10 a slow-moving thunderstorm began dumping torrential rain onto the Winona streets. After ten hours the Mississippi River overflowed its banks, sweeping away bridges, barns and livestock. The entire south side of the city was under water, and people had to be removed on rafts floated down the debris-laden streets. For three days the storm lingered, dumping torrential rain on the city, flooding the sewers and flooding basements and floors everywhere. Temporary walkways had to be built to allow people to pass through the streets so they could go to the aid of neighbors and family. On the second day of the flood, Henry Schultz was struck by lightning and killed. In all the storm caused in excess of $200,000 in damages and became one of the worst storms in the state's history. The Humasons, with a newborn and a sickly father at home, had to depend on the kindness of their friends and family in getting the food and supplies they needed to weather the deadly storm.

The long work of putting the town back together began shortly after the siege ended. As the floodwaters subsided and the river sank back to its normal depth buildings had to be cleared of their furnishings and their walls aired out and repaired. Bridges, swept away in a torrent of river water, had to be rebuilt so train service could resume. Debris was brought to the riverfront for removal by boat downriver. The townspeople worked together to see that everyone who needed help got it. People like William Humason, beleaguered from his bout with typhoid, tried to lend a hand but were too weak to be much good. The cool, damp weather sticking to William's bones leached the energy from his frail body.

As summer began and the weather warmed William's stamina and his spirits finally lifted, and the whole family breathed a sigh of relief. By late August he was feeling much more himself again as Milton celebrated his ninth birthday. They had eluded tragedy, and life, it seemed, was slowly getting back to normal.

As the clock struck midnight on New Year's Eve and the Church bells chimed ringing in the twentieth century, the Humasons joined with their neighbors in celebration of the occasion. Although most people remained cautious, there was plenty for the young nation to celebrate. Nearly 2 years earlier the country had gone to war against the once mighty Spanish military and won a decisive victory. Although much of the country was against the imperialist movements of the McKinley government in the year since the war, victory and a recovering economy helped to ease these frustrations. As the twentieth century dawned, the United States had established itself as an emerging economic and military power, and McKinley was being hailed as the man who had engineered it. The country was awash in the finest inventions the world had to offer and new inventions seemed to lurk around every corner. The earthen grit of the old world was being steadily replaced by the clamor of the age of iron and steel. Growing up in this ever-changing world, Milton would still always have one foot planted firmly in the old one, a fact that in the years to come would test his courage, perseverance and guile.

Chapter 2
1902

Abstract Family tragedy and the lingering symptoms caused by William Humason's bout with typhoid, lead the Humason family to move west to Los Angeles where they meet Laura Humason's sister, Alice, and her husband, Henry Witmer. United in an effort to overcome their misfortune the families begin to build on the future Henry Witmer and his brother's had begun decades before.

On April 22, 1899, Laura Humason received a telegram from Los Angeles. Her brother-in-law, Henry Witmer, had sent the message to tell her that her sister, Alice, had given birth to a boy. Joseph Petterson Witmer was born on April 21, 1899, in Los Angeles, California, the first and only child of Henry and Alice Witmer. With William just home from a Chicago hospital, the comforting news of her sister's successful birth came as further proof that the family's fortunes were rising again.

For years, Laura had been reading letters from Alice telling her about a man she had met and their ongoing courtship. Alice Petterson had met Henry Clayton Witmer in 1894 while on holiday in Lake Tahoe. Witmer, a well-known banker and real estate developer in Los Angeles had come to the country for some hiking and fishing. Fourteen years her senior, the older man was quite taken with Alice's charm and beauty and the two began a sometimes stormy, four-year courtship shortly after, mostly by mail. Henry had filled his letters with bits of information about his various business interests in banking, real estate and oil. He often sent oranges from his orchard with little notes hinting at his intentions. Alice turned him down twice before she finally agreed to marry Henry, but at last he persuaded her in the early part of 1897.

H.C. Witmer and his younger brother, Joseph, had made a name for themselves from the moment they set foot in Los Angeles. The two had established the Witmer Brothers Company in their childhood home of Monroe, Wisconsin, in 1882 and founded the Citizen's Bank of Monroe in 1883. In 1884, the Witmers arrived in Los Angeles with their sisters and their widowed mother. The oldest of the Witmer children, Mary, was married to a Civil War veteran named Sam Lewis, who joined Henry and Joe in operating the family business. Seizing on the opportunity to provide transportation to the wealthy residents living in the hills east of the city, Witmer Brothers created the first cable car service in the city in 1885. The cable cars sat on thin rails and were pulled along by a cable buried in the street. The driver

© Springer Science+Business Media New York 2016
R.L. Voller, *The Muleskinner and the Stars*,
Springer Biographies, DOI 10.1007/978-1-4939-2880-4_2

controlled the movement by operating a brake that grabbed or let go of the cable to move or stop the car. Another brake was used to hold the car in place when it was stopped on a hill.

The cable car was a big success, and the brothers made sure the line ran near their home, regularly using it to commute to and from town. Riding this success the Witmer Brothers Company opened the California Bank in 1887. The bank operated successfully until 1903 when the Witmers merged it with another local bank, changing the name to the American National Bank. Witmer Brothers also cashed in on the land boom that was peaking during this period, purchasing some 650 acres of land in the Crown Hill area, which offered a beautifully picturesque view of the city. They created the Los Angeles Improvement Company to manage their real estate and oil (discovered on their land in 1892) interests and built spacious homes in Crown Hill for their families. Henry Witmer's home stood at 1422 West Third Street, while Joseph built his home next door. Sam and Mary Lewis lived across the street at 1425 West Third Street.

They also bought land in Lordsburg (La Verne), where they built a citrus farm with stables and a large orchard. The orange groves provided plenty of income and the farm offered a home away from home for the family with plenty of hiking, fishing, horseback riding and picnicking. The family named the farm "Aljoleto" after the Witmer women: Agnes Lewis, Josephine (Mrs. Joseph Witmer), Letha (Lewis, daughter of Sam and Agnes Lewis) and Torrie (Anna Victoria Witmer), but Henry was fond of saying that the name meant "nothing better."

By the time he met the beautiful 24-year-old Alice Petterson, Clate, as he was called by family and friends, was one of the best known men in Los Angeles. He had an infectious personality that, combined with his driving ambition, made him a formidable figure among local businessmen.

In the fall of 1897, with the wedding just months away, Joseph Myer Witmer died suddenly of heart failure at the age of 39. The shock of the loss was almost unbearable for Clate, who had not only lost his brother but his best friend and business partner. The younger Witmer brother left behind his wife, Josephine Witmer, and three children, Mary, William and David. Inconsolable, Josephine left the city with the children and returned to her family in Massachusetts. After the tragedy Clate had thrown himself into the family business. In an attempt to cover his late brother's duties to the firm, he enlisted the services of his brother-in-law, Sam Lewis, asking his aging partner to help take on some of the added burden. In January of 1898 the situation worsened when Sam Lewis also died suddenly of heart failure. The war veteran, father and husband was 56 years old and thought to be in perfect health. In the space of six months the Witmer's had lost two heads of household and left Henry Witmer without his two most trusted business partners. Reeling from the loss Henry sold his brother's home, while his sister, Mary Agnes Witmer, her daughter, Letha, and son, Ralph, stayed on in their family home on West Third. Anna Victoria, who had never married and was well-schooled, helped out filling in as office manager while keeping the books.

By June life had finally settled down again. So, in a bittersweet ceremony that was shadowed by the tragedies of the past year, Henry and Alice had married in a private ceremony. It hadn't taken them long to get started on having a family either. Milton likely met the joy and mystery of his mother's stories about his aunt out west with the kind of unsubtle apathy that only a 9-year-old boy with fishing on his mind could produce. There might, in fact, be life elsewhere, but who cared? For Laura Humason and her sister, Alice Petterson Witmer, sharing in the births of their newborn daughter and son was the best way for the sisters to begin the new century.

* * * * * * * * * * * * * * * * * * * **

In the election of 1900 William McKinley was swept into his second term as president by the largest margin in history. Riding a wave of public adoration for ending the recent depression McKinley had continued the open-door policy that had helped make him popular among Republicans and Democrats alike. The mild-mannered and self-effacing McKinley had chosen Theodore Roosevelt to be his new vice president, a man who had public adoration mounted on the wall in his trophy room. As Assistant Secretary of the Navy during the war with Spain, Roosevelt had acted rashly and had to be checked as McKinley maneuvered U.S. forces and curried public support for an attack on the Spanish military. Nevertheless, the young adventurer had garnered attention when his band of South Dakota ranchers, Indians and New York City policemen, known as the Rough Riders, rode to victory in the Battle of San Juan Hill in Cuba. Small of frame but large of stature, Roosevelt defied the social norms of his time. He was a devout family man and could often be found roughhousing with his children instead of concentrating on work.

McKinley was sworn into his second term on March 4, 1901, outside the Capital building. Although he was lauded for his efforts in getting the U.S. economy running again, he had suffered greatly in the public eye for his handling of foreign affairs. Just as McKinley had determined to keep troops in the Philippines after the war an anti-imperialism movement took hold across the country. The leader of the movement was the 65-year-old steel magnate, Andrew Carnegie. Although he was a lifelong rabid Republican, Carnegie had always been against the country's involvement in foreign wars, and his otherwise good feeling about McKinley was shaken when the latter decided to annex the tiny island nation in the South Pacific. When faced with a choice between McKinley and William Jennings Bryan, the Democratic candidate who had staked his election on the importance of coining silver instead of relying solely on gold, Carnegie decided to stay with the Republican ticket.

Andrew Carnegie had been born in Dunfermline, Scotland, in 1835, and moved to America with his family in the 1840s, where he subsequently made a small fortune during the Civil War from speculations in oil, iron and rail. A shrewd and ruthless businessman, Carnegie's central philosophy was, "Put all your eggs in one basket and then watch the basket," which he did to an almost unparalleled degree. Throughout the latter half of the nineteenth century, Carnegie had built up his steel manufacturing plants to be the biggest and most efficient in the world.

In March of 1901, a consortium of the top business leaders led by J.P. Morgan offered Carnegie the sale of his Carnegie Steel Works for the price of $480 million.

After briefly considering the deal, Carnegie accepted it, which made him one the richest men in the world, behind only John D. Rockefeller. Suddenly flush with capital Carnegie was faced with a new challenge, one that he had been cultivating for more than half his life. In Carnegie's view, a man should spend the first third of his life getting as much education he could, the second third making as much money as he could, and the third giving as much of his money back to the community as he could. He had summed up his thoughts on the subject in his 1889 publication *The Gospel of Wealth,* which had become a popular read for anyone seeking his fortune in the business world.

In keeping with his commitment to redistribute his wealth to the community as responsibly as possible, Carnegie was planning an institution in his name to aid in the development of the most gifted and talented people the country had to offer. He was in the midst of planning for his new organization when President William McKinley was shot, twice in the chest, by an anarchist named Leon Czolgosz while receiving guests in the Temple of Music at the Pan-American Exposition in Buffalo on September 6. One of the bullets lodged in McKinley's stomach, and the president died eight days later from septicemia. A disgruntled Ohio steel worker, Czolgosz would later say he killed McKinley because he was, "the enemy of the good people—the good working people." He said this as he was being strapped to the electric chair.

On September 14, with the country mourning the loss of their greatest leader since Abraham Lincoln, Vice President Theodore Roosevelt was sworn in as the new head of government and the Republican party. About McKinley's assassin Roosevelt would simply say, "When compared with the suppression of anarchy, every other question sinks into insignificance." Long a temperate man with a long streak of philosophical reserve, Carnegie viewed the young president as a "dangerous man," and feared Roosevelt might be too rash a thinker to continue in the footsteps of his predecessor. Little did he know the new president would eventually reign as one of the great presidents in U.S. history.

At this time Carnegie was putting the finishing touches on his new philanthropic organization, The Carnegie Institution of Washington, and it was his desire to give the government the sum of $10 million in gold bonds to establish it as a national trust. Seeking to avoid public fallout for accepting such an overwhelmingly large sum from a man that so much of the public regarded as a robber baron, Roosevelt politely turned down Carnegie's offer, but accepted his proposal to join the institution as one of its board of trustees.

Understanding and accepting the president's wishes, Carnegie quietly finalized and incorporated The Carnegie Institution on January 4, 1902, with an endowment of $10 million. The first meeting of the nascent institution's prestigious board of trustees, which included President Theodore Roosevelt, was held January 29, 1902.

* * * * * * * * * * * * * * * * * **

Laura Humason was thinking her family needed a lift. Faced with another rough Minnesota winter, the California native was nearing the breaking point. William's overall health had improved, and he was in good spirits during the warmer months, but summers were short-lived in the northern Midwest of the country, and fall and winter wreaked havoc on his condition. Furthermore, her sister Alice had been sending her letters consistently since her marriage to the millionaire banker, Henry Witmer in Los Angeles, imploring her to move the family out there so they could be together. William had resisted the notion because of his father's health, but the situation in Los Angeles was sounding more and more dire. Alice was concerned for her husband's well-being. In the years since the death of his brother and brother-in-law, Clate had been overwhelmed by the details of the ongoing operation of the family business, and he needed help from able-bodied and trustworthy men to help secure the future of the business. A move west would be good for William and improve the Humason family's fortunes. They would probably not be afforded the same opportunity if they stayed in Winona. William and Laura had grown up in the northern part of the state, so going back would be a sort of homecoming for them. Knowing the southern California air would boost his constitution, William finally made the decision to leave his father and mother in the care of his family and move his own family west.

The news wasn't good for 10-year-old Milton and his brother Lewis. The brothers had been inseparable since early childhood and the town of Winona had been their playground. Together they had crawled through every crevice, climbed every hilltop and fished every inch of the rivers and lakes within walking distance of their home at 319 Center Street. Faced with the reality that they would soon be leaving their hometown they tried to imagine the wonders that awaited them on the other side of the great railway between Winona and Los Angeles. It was this potential that Milton was imagining as he stood on the wooden train platform at the depot in town with his family. Soon the train from the Illinois Central rolled in on the southbound track. Appearing at the foot of the stairs the conductor called out, "All aboard," and excited travelers began to disappear into the cars along the platform. Soon, with a long whistle and a start, the train slowly rolled out of the station and down the tracks to new adventures. When they arrived at Union Station in Chicago, the Humasons boarded the Union Pacific Railroad bound for San Francisco. The old line was the original route cut through the heart of the central plains.

On the train Milton and Lewis saw for the first time the wonders of the country west of Minnesota. The train rolled through the Rocky Mountains on its way to the Great Salt Lake in Utah and later made its way through Donner Pass, named for the ill-fated party of settlers who were stranded there in the snow over fifty years earlier. (Trapped without food, the party had resorted to cannibalizing its dead to keep the survivors alive.) The family lingered in San Francisco for a time to visit family and friends and see the town before loading onto the Southern Pacific bound for Los Angeles and their new life on the shore of the Pacific Ocean.

* * * * * * * * * * * * * * * * * * **

On January 10, 1902, George Ellery Hale was sitting at home in Williams Bay, Wisconsin, reading the *Chicago Tribune* when he ran across a headline that caught his attention. Andrew Carnegie was giving $10 million to an institution bearing his name. The Carnegie Institution of Washington, he read, had been established had been established by the millionaire steel magnate turned philanthropist "to encourage investigation, research and discovery to the broadest and most liberal manner, and the application of knowledge to the improvement of mankind." Stunned by the news, Hale rocked back in his chair. This might be the break he was looking for.

Born in Chicago in 1868, Hale had grown up in the lap of luxury. His father, William Hale, had made a fortune installing an elevator he had patented just prior to the Chicago Fire of 1871 in the many high rises that sprang up in the city in the fire's wake. Sickly as a child, Hale made up for his chronic health issues with curiosity and enthusiasm, especially in the area of astronomy and the emerging field of astrophysics. From an early age Hale's doting father had supported his inquisitive nature and staked his astronomical investigations with lavish gifts. The younger Hale made the most of his opportunities. While in college, at the newly formed Massachusetts Institute of Technology, he had invented a spectroheliograph, intended for close study of his favorite star, the Sun. Hale based his senior thesis on the new instrument and planned to develop it as investigations revealed necessary improvements to the design.

In June of 1891, Hale dedicated his first observatory, founded with his father's support on the grounds of the family's home in the well-to-do area of Hyde Park in Chicago. The Kenwood Observatory, as it was known, consisted of a single observing dome and a 12-inch Clark telescope. By then, at the age of 23, Hale had already started to make his presence felt in the world of astronomy. The *Astrophysical Journal*, which began publication later that year, was one of Hale's lasting creations of the era. His new observatory had been designed by D.H. Burnham, one of the world's leading architects and a good friend of William Hale.

Burnham would gain worldwide attention two years later when the Columbian Exposition of 1893, of which he had been the chief designer, opened to broad acclaim. The fair offered George Hale a chance to show off his latest creation, the 40-inch refracting telescope, the largest refracting telescope ever built. With the blessing of the University of Chicago, Hale had convinced the railroad magnate, Charles Yerkes, to agree to fund the telescope and observatory in 1892. Only the 60-foot tube, 40-foot-high pier and state-of-the-art electronic and fully automated rotation system were on display during the fair. On dedication in 1897, however, the Yerkes telescope officially surpassed the 36-inch refractor, at the Lick Observatory near San Francisco, as the largest operating telescope in the world. The giant was housed inside the 90-foot-diameter dome and was the first large telescope whose movement was controlled electronically. Like the Great Lick Refractor, the Yerkes refractor featured an observing platform that was raised and lowered by means of motorized cables. Initially the observing floor had been the center point for misadventure. Early in the morning on May 29, 1897, the 75-foot-diameter floor, weighing almost 38 tons, collapsed into a pile of rubble at the bottom of the dome. Fortunately for Hale and the others working at the observatory they worked at night and were sound asleep in their beds when the observing floor came crashing down. By October of the same year the floor had been rebuilt and the telescope was officially dedicated in a lavish ceremony on the observatory grounds.

Not really a "less is more" kind of guy, Hale had already begun dreaming of larger telescopes to be used in new and more sophisticated facilities for the sole purpose of studying and reporting on problems concerning stellar evolution. Hale and his lead telescope designer and technician, George Ritchey, had been experimenting with large reflecting telescopes for some time and Ritchey had built a 24-inch reflector at Yerkes. In testing the new telescope design proved to have as good or greater observing power than the 40-inch refractor, at a fraction of the size of both the primary objective and tube. While speaking with his father one evening, Hale began to ruminate about a reflector with a 5-foot primary mirror. A telescope of that size, he reasoned aloud, could revolutionize our knowledge of the universe. So convinced was William Hale of the veracity of his son's claim that he promptly ordered a 60-inch glass blank from the French Plate Glass Company in St. Gobain, France. With the new disc in hand, Hale began dreaming of a location suitable for a new telescope, incredibly technically advanced, with a massive primary mirror and never before seen light-gathering capability. The location must be perfect for such an instrument to be used to its fullest potential.

Four years later George Hale was still dreaming of the moment when his new telescope would come online. To succeed he would need a suitable location for an observatory. The winters in Williams Bay were cold and dreary, and the number of good seeing days were limited. To maximize the superior viewing power of his new telescope, Hale wanted to find a place with a climate conducive to consistent day and night time observing. He had heard of an expedition to Mount Wilson in Los Angeles, California, by the Harvard Observatory in January of 1889. Although the climate was up to the standards necessary for an observatory, Harvard had abandoned the idea of building a new facility on the mountain due to the difficulty in getting supplies up the steep trail to its summit. Always a pioneering spirit, Hale had no such reservation. If the climate and seeing proved ideal, he knew he could successfully build an observatory at Mount Wilson. The bigger issue was funding.

In 1901, Hale had tried to get Chicago millionaire John D. Rockefeller to fund the project. The 33-year-old Hale, invited Rockefeller to Williams Bay to view the Sun through a spectroheliograph attached to the Yerkes refractor. Although impressed by the display, Rockefeller declined to help. It was a disappointing blow to Hale, who was convinced after his visit that the wealthy oil man would consent to funding the new facility. Consumed by the work of getting the Yerkes observatory running smoothly and efficiently, Hale had had little time to devote to finding the right resource to fund his latest pet project.

Now as he sat in his study, sipping coffee and gazing at the snow falling outside his window that cold January morning, Hale could see his vision coming into focus. One of the goals of Carnegie's trust was to "discover the exceptional man in every department of study, whenever and wherever found, and enable him, by financial aid, to make the work for which he seemed especially designed his life work." The paper was indulging a bit on the actual language of the trust, but it had the general gist, and Hale couldn't be happier to read it. At last he had found a source of funding for his observatory. An organization with an endowment the size of the Carnegie Institution of Washington, dedicated as it was to scientific research, would be capable of funding the development of an entire mountain observatory. All Hale had to do was convince the Carnegie board of trustees that his idea was both viable and scientifically worthy.

Carnegie was aware of the potential of a well-designed mountain observatory, having written on the subject in his landmark article, "The Gospel of Wealth," in 1889. Calling attention to the Lick Observatory Carnegie wrote, "If any millionaire be interested in the ennobling study of astronomy—here is an example that could well be followed…" Having read this article, Hale had reason to believe Carnegie might be interested in funding a new observatory like the Lick. He also knew that a giant reflecting telescope with a 5-foot-diameter primary mirror would have awesome light-gathering potential, and he had a hunch the old Harvard site on Mount Wilson would make a suitable location to house it.

Summoning his formidable instincts and knowledge in such endeavors, Hale wrote a letter to the Carnegie board's vice-chairman, Dr. John Billings, a noted surgeon and director of the New York Public Library. In his letter Hale spoke of the glorious opportunity that would be afforded the science community if an observatory, built in a suitable location with a one-of-a-kind reflecting telescope housed in it, could be established. He included photographs of the Orion and Andromeda nebulae, taken with both the Yerkes refractor and the new 24-inch reflector built by Ritchey at the Yerkes lab, showing clearly the superior resolving power of the smaller reflector. In closing, Hale asked Billings for assistance in bringing the matter before the CIW committee. Hale's reputation and achievements as an innovator, scientist and observatory builder no doubt preceded him. At age 34 he had already founded two observatories, built the world's largest refracting telescope and invented a device that enhanced astronomers' ability to study the inner workings of the Sun, not to mention starting the *Astrophysical Journal* and other publications. After considering the matter, the Carnegie board agreed to funding an expedition to find suitable locations for an observatory. Hale was chosen to lead one of the legs of the expedition and immediately started making plans for a visit to the Los Angeles area and a hike up the steep mountain trail to the summit of Wilson's Peak.

* * * * * * * * * * * * * * * * * * * **

If Milton had seen an automobile before, back home in Minnesota, it would have been a novelty and not much more. The cantankerous contraptions weren't numerous there before 1900, and they could never take him into the woods along the trails where he wanted to go, so what good were they, anyway?

In turn of the century Los Angeles, however, automobiles were everywhere. As a younger boy he had heard stories about the great cities of America such as New York, Chicago and Minneapolis/St Paul, but now he was witnessing one first hand. Although Los Angeles wasn't large by those city's standards, it was one of the nation's fastest growing cities and had a population five times the size of Winona. The clamor and bustle of the city's downtown area was overwhelming, wired from one street to the next with telephone cables, electric street lights and street cars. Throngs of people walked the streets shopping the awning-crowned storefronts that were too numerous to count or comprehend. Horse drawn carriages competed with automobiles, trolley cars and bicyclists for right of way on the busy rutted streets.

Everywhere Milton looked a group of suited men or women dressed in the long Victorian-era skirts and large-brimmed hats could be seen walking or talking. Boys ran between the vehicles on the streets or sold newspapers on the corners at busy intersections. The streets, lined with shops selling everything from hosiery to haircuts to horseshoes, were crawling with customers. Parks and flower beds seemed to spring up at a moment's notice, and orchards lined the countryside just outside the city limits. The Los Angeles Produce Market, with rows of horse-drawn carts, teemed with grocery shoppers at its location in a vacant lot at 9th and Los Angeles Streets. The Angel's Flight Inclined Railway ran passengers from Hill Street for two blocks up the steep hill to its Olive Street terminus, where sightseers could then climb an observation tower for a view of the city and its surroundings. Steamships and paddleboats lined the shore of the Pacific Ocean, which stretched into the horizon until Milton could actually see the slight curvature of Earth. Oil derricks lined the streets and houses at the southern end of town, stretching west to east all the way to the foot of the San Gabriel Mountains.

As the train from the Southern Pacific Railroad steamed into the station at Santa Monica and slowly came to a halt, Milton sat beside his parents trying to process the scene. The conductor interrupted his daydreaming, shouting out the station name. The family gathered their things and stepped down the steep steps to the railway platform where the Witmers, Alice and Henry and their 2-year-old son, Joseph, were waiting for them. With their arrival, the sisters, Laura and Alice, were united again, and it couldn't have come at a better time. On the way out of town they stopped by the California Bank Building on 2nd Street and Broadway, where the Witmer brother's business had been born in 1887, and rolled past the family-owned oil derricks and other land holdings. As they worked their way out of town Henry Witmer described in detail the growth in the city over the past twenty years in vivid detail. An energetic and naturally good-spirited man, Milton liked his uncle from the moment they met.

The two families soon made their way out of town on 2nd Street to the Witmer's home in Crown Hill, a remote area inhabited by several of the city's richest families. The San Gabriel Mountains loomed on the eastern horizon as they made their way, and Milton could see the Pacific Ocean stretching west as he looked behind him. After purchasing 650 acres of land in the area, the Witmer brothers and their brother-in-law, Samuel Lewis, had each built a large home on Third Street on a hilltop surrounded rolling hills. Henry Witmer's home was at 1422 Third Street. After his untimely death in 1897, Joseph's widow, Josephine, had sold the house at 1400 Third Street with Henry's help, and moved with her three children, Mary, William and David, back to her childhood home in Massachusetts. Henry's sister, Mary Agnes, had remained in the house at 1425 Third Street following the death of her husband, Samuel Lewis. Henry Witmer had taken on the responsibility of heading up both homes, along with his wife and son, his sister and her two children, Letha and Ralph Lewis. The 2nd Street Park was built near a wooded area by Henry to provide a place for local residents to enjoy an afternoon outing. The park featured a pond for boating at its center and a playground with swings, a seesaw and a sandbox for the children. The Witmer's large home was more than sufficient to house the two families while the Humasons looked for a suitable home for themselves.

When the family arrived at the Witmer home, they were greeted by Mary Lewis, her daughter, Letha, and her son, Ralph. Letha Lewis had been born in 1869 in Monroe, Wisconsin, where the Witmers and Lewis's had lived prior to their move west, shortly after the transcontinental railroad was completed. Beautiful and well-schooled, Letha had a penchant for the theatrical and a willful independence that was captivating in her youth but made her seem more eccentric as she got older. Ralph Witmer Lewis was twelve years younger than his sister and was attending military school at the time. His father, an officer in the Union army during the Civil War, believed in strict obedience and respect for authority and made sure Ralph understood the necessity and honor of serving his country. Although they were ten years apart in age, Milton and Ralph would remain lifelong friends.

In the course of history 1902 would become a pivotal year for Milton and his kin. As they settled into their new life within the bosom of the Petterson/Witmer side of the family, their new surroundings provided a much-needed sense of hope and security for all.

Meanwhile, on the other side of the country, George Ellery Hale made plans for his own visit to the area and the summit of Mount Wilson. The 34-year-old observatory director was working under the auspices of the newly formed Carnegie Institution of Washington and its charter, "to discover the exceptional man… whenever and wherever found…" Little did anyone know at the time it would be Milton Humason, as much as any man, who would come to personify that edict in the decades to come.

Chapter 3
Mount Wilson

Abstract On his first visit to Mount Wilson for summer camp, Milton Humason is introduced to the magic and mystery of the mountain wilderness. One of the many attractions of the mountain is the presence of George Ellery Hale who is taking his first tentative steps toward building an observatory on the summit. After visiting the mountain again the following year for summer camp, Milton decides to quit school to take a job at the recently established Mount Wilson Hotel. In the next five years he becomes a mule driver, leading teams of mules hauling supplies up the steep mountain trail, and meets the woman of his dreams, Helen Dowd.

Winter in southern California was nothing like winter in Minnesota. Milton stood in shirtsleeves outside a fenced-in corral watching his uncle break in one of the mares on the family farm in Lordsburg. It was a warm and sunny afternoon, and he had never known what it felt like to be warm during the coldest months of the year. A person could get used to this. Uncle Clate, as Milton was beginning to refer to his uncle, had purchased the land and built the ranch in the early 1890s with his brother. The family had nicknamed the ranch, Aljoleto, forming the name out of the letters in the names of the four (at the time) women in the family, Agnes Lewis, Josephine, Letha (Lewis) and Torrie Witmer, Clate's older sister. The ranch, with its vast orange groves, riding trails and ravines, fishing and swimming holes, was a favorite place to visit for Milton and Lewis. On weekends the family would often meet there with friends for hiking and cookouts and sing-alongs by the campfire.

The dry California air and mild climate had an immediate effect on William Humason's health. Months after their arrival he was feeling more himself than he had in years. This was good news to the ladies of the house, Laura and Alice, who had been through so much handwringing in the past years. As William got to understand more of the totality of the family's holdings, he could be of more use at home and at the Witmer Brothers' offices. Alice and Henry were insistent that the Humasons stay with them in their big house in the hills, but William was adamant that he carry his share of the load and that he would find a home for his family as soon as possible.

The world had been full of surprises for Milton since his arrival on the west coast. For one thing he had never heard a telephone ring inside his house before he

© Springer Science+Business Media New York 2016
R.L. Voller, *The Muleskinner and the Stars*,
Springer Biographies, DOI 10.1007/978-1-4939-2880-4_3

moved in with his aunt and uncle, but now one rang fairly often as calls came in for Henry Witmer on various business matters. He had never seen a motion picture before, either, but that year *The Great Train Robbery*, the first silent film to reach mass audiences, was introduced at Tally's Electric Theater. The movie had been filmed and edited by Thomas Edison at his studios in West Orange, New Jersey. The first World Series of Baseball was played in 1903, as well, and the family's holding company sponsored a local baseball team. Milton and his brother would accompany their uncle and father on trips to the ballpark to watch the games.

During the summer of 1903, Milton Humason went to summer camp on Mount Wilson. The mountain was named for Benjamin Davis (Don Benito) Wilson, a rugged mountain man and veteran of the Mexican-American War, who had made his way west in the 1840s from Tennessee and settled in the Los Angeles area. There he made a fortune in everything from merchandising to agriculture to real estate, and later became mayor. When he died Benjamin Wilson was one of the wealthiest and most respected men in California. He was also the grandfather of General George S. Patton, who would become famous leading the U.S. Third Army after the invasion of Normandy during World War II.

The 6000-foot peak was reachable by either of two paths that slowly wound their way up through 8 miles of often treacherous terrain. The old Indian trail led from Sierra Madre up the canyon near the Little Santa Anita, and the new trail, built by the Pasadena and Mount Wilson Toll Road Company, led from Altadena up the southern slope of the mountain. Neither trail was wider than 2 or 3 feet at any point, and footing was often treacherous for both man and beast.

Near the summit stood an old log cabin known as the "casino" that had been built in 1893. There were two camps on the mountain as well. Martin's camp was located on a ridge between Mount Wilson and Mount Harvard, and Strain's Camp was on the north side of the mountain. Strains Camp was located near a spring and so had a good supply of fresh water. The camp, open only during summer months, was located in a cedar grove, making it an ideal spot for hikers and campers.

For 12-year-old Milton the trailhead of the new "toll road" was as intimidating as it was exciting. His bright blue eyes must have darted from scene to scene as he stood in the boarding area, his tousled, wavy brown hair tucked under a slightly worn Mounty-style hat. A small stable stood at one end of the grounds, and a number of shabby-looking mules stood in a small fenced-in corral at its side. On the other side of the stable house stood a storage area where short posts jutting out from the walls held saddles and bridles. On the other wall of the small holding area shelves and drawers held mane combs, shoeing equipment and other supplies. Saddled horses stood at a hitching post near the stable, and smoke rose above the clanging of steel at the blacksmith's shed. The morning sky was veiled in a layer of low-lying clouds that shrouded the summit of the mountain.

A man standing near a small wooden hut barked orders at a spirited group of young muleskinners who were busy readying mules for the journey up the steep mountain trail. A sign on the hut read Bassett and Son, and another, nailed to the side of the hut, listed the prices. The cost of renting a mule for the day on the trail was 15 cents. After conferring with the barking man for a moment, Milton's father paid him, and the order was shouted out to secure Milton a mule. The cowboys working the grounds piqued Milton's interest more than anything else. Chewing tobacco the rugged, sharp-tongued and dusty young men cursed their charges, who groaned in displeasure at having to bear the loads of supplies being set on their backs. Watching the enterprise that morning was a revelation to Milton. After months of touring banks, oil fields and plots of land here was a job he could understand.

When the mules were loaded and the train was ready to depart, the cowboys mounted their horses and organized their party, one each at front and rear of the train, and a couple interspersed throughout to help out in case of emergency. The trail was very steep and narrow at some places, and a turn taken too wide would result in the mule and its payload toppling over the edge. After a snowstorm one winter a mule slipped off the edge of the trail and fell down the mountainside, landing in shoulder-deep snow. The snow was too deep to haul the beast to safety so, instead, a cowboy had to walk food down to it every day until the snow melted sufficiently to help it back up to the trail.

After what seemed like an eternity the call was given to move out, and Milton waved goodbye to his dad. The mood changed as the long train made its way up the mountain. The morning light beneath the clouds was gray as Milton sat in the saddle glancing down at the twitching ears of his burro. The ride up was not for the faint of heart as the trail narrowed at points, exposing steep ravines and rocky cliffs. The mountain side was brown with dirt, red clay, stone, yucca and sage. Tall ponderosa pines reached into the skyline, and vultures circled ominously overhead as he rode. His charge, a surefooted young mule, walked along as if it was out for a leisurely stroll and didn't have a care in the world. If the beast knew it was busy hauling Milton up the side of the mountain it showed no sign of it, stopping at intervals to nibble at grasses and shrubs. Another mule, with utter disregard for its rider, decided to lay down for a nap, and had to be coaxed up by a cajoling cowboy. Milton could certainly understand their impatience, but he noticed that while they were quick to curse an animal for bad behavior, the muleskinners seemed to have a kind of kinship with the mules. They took pride in knowing each mule's unique traits, and joked among themselves about them, the riders and their boss with the kind of recklessness that only a free man could possess. There was a camaraderie among them, a code of honor, respect and brotherhood.

As the day wore on, the long line of hikers, campers, mule teams and horses approached the thin veil of haze that had hung over their party like a gray blanket all morning. Milton himself disappeared as he rode slowly into the fog. The visibility was only a few feet as he made his way warily along the trail, sure that each next step his mule took would be his last. It was the kind of climb that made the hair stand up on the back of his neck.

For what seemed like hours the fog blanketed the party ascending the mountain. Finally, they emerged from the darkness, and, all at once, the hazy early morning gloom was replaced by a beautiful sunlit sky.

Not long after they had emerged from the fog the party ambled into Strain's Camp, a sprawling campground with a number of white canvas tents stretched across an open area in a forest of cedar pines. Thin ropes stretched from the corners and sides of the tents to the ground, where stakes held the tent walls steady and stabilized them against wind and rain. Fire pits spotted the grounds, some of them still smoldering from the morning's activities.

The beauty and danger of the wilderness surrounding him was captivating. During his stay at camp he spent his days fishing and hiking, shooting at the rifle range and learning important survival techniques. Rattlesnakes were rife on the mountain, and bear and mountain lion roamed freely. A branch of the San Gabriel River down the trail provided a water hole where Milton and his friends could spend an afternoon swimming and leaping off the warm rock faces into the cold mountain water. Camp instructors helped them recognize and understand the plantlife and wildlife of the forest. Near the summit there was an old rundown cabin that became a prop for cowboys and Indians. Milton sought any occasion to ham it up with his friends, a trait that would stay with him his entire life.

By the end of his first stay the only thing Milton could think of was making the mountain his permanent home. He had no idea how to make that happen, but, as he made his way back down the mountain to his life in the sleepy hills outside Los Angeles, Milton made up his mind that he was going to try.

* * * * * * * * * * * * * * * * * * **

In June of 1903, George Ellery Hale arrived in Pasadena on an expedition to seek a suitable spot for his next observatory. Observatory building was becoming an obsession for Hale, and with good reason. The brilliant young scientist had a knack for finding interest in as well as the money for his ideas. Yerkes Observatory, from which he was on a sustained leave of absence, was the second observatory he had created in a decade's time.

At the station Hale was joined by W.W. Campbell of the Lick Observatory, and the two set out for Mount Wilson. As director of the leading observatory in the country, located in the mountains south of San Francisco, Campbell had been chosen as one of the members of the research team charged with finding a suitable location for a large observatory, to be built with funding from the Carnegie Institution of Washington. Campbell had good knowledge of the area and was eager to meet with Hale and explore Mount Wilson for himself. Hale was good friends with Campbell's predecessor at the Lick, James Keeler, and the expedition up the mountain would give him a chance to size up the Yerkes director. Already well known for his ability as an instrument designer and creator at M.I.T. and for founding the Kenwood and Yerkes observatories, Hale was in the unique position of owning the enormous glass blank his father had ordered years before from St. Gobain in France. Producing a single piece of glass of that size was no mean feat and required considerable knowledge, skill and money to complete. That was why William Hale had sought the expertise of the centuries-old glassworks across the Atlantic.

Having the enormous disk in his possession, and his track record in founding observatories, definitely made Hale the leading candidate for founding one under the Carnegie trust. Successfully building the greatest observatory in the world on the summit of an untamed and unforgiving mountain wilderness would require a mountain of cash, and Hale knew it. William Hale, who died of Bright's Disease in 1898, had left behind a small fortune, but Hale knew it would not be enough to ensure success in California. Attempting to build his new telescope and observatory using his family's fortune would simply exhaust his father's legacy and would likely end in failure. When Andrew Carnegie announced the founding of his institution the year before Hale thought he had found the resource he needed. It was a grand gesture by the steel magnate turned philanthropist, and Hale planned to do his part to fulfill the institution's promise. He was certain he could convince the board at the Carnegie that a mountain observatory with a huge new reflecting telescope was worthy of the institution's funding.

Riding burros along the mountain trail, Hale and Campbell talked about the seeing conditions at their two observatories. Although equipped with the world's largest refractor, the seeing conditions at Yerkes left the astronomers there wanting for air and weather conditions that were more conducive to quality and consistent observing time. The observatory's location near the western edge of Lake Michigan made for long winters and hot, humid summers. More importantly, the surface conditions in the area created turbulence that distorted the image of the stars they were studying. Hale suspected that the seeing at a seaside location high on the windward edge of a mountain top would draw cool air from the sea upslope, sweeping away the warm air and settling the turbulence caused by warm surface temperatures. Campbell confirmed that the seeing at the Lick Observatory on Mount Hamilton was far superior to that at other facilities. The newly appointed Lick director was a native Mid-westerner who had studied at the University of Michigan and later returned there as an instructor in astronomy.

They had talked about the expedition from Harvard to study seeing conditions on Mount Wilson years before. Hale was at the university the year the expedition was undertaken and remembered looking with high expectations. In the end the university had decided the rattlesnake-infested site was too remote and that building a facility there would create costly time delays and other unforeseen challenges.

The day on the trail united the two observatory directors in the pursuit of scientific research. Large telescopes like the Yerkes refractor and the 36-inch at the Lick were beginning to get a better picture of distant stars and other phenomena. On their trip Hale and Campbell agreed to bring their institutions together toward common goals whenever possible.

At the summit they were greeted by W.J. Hussey, an astronomer at the Lick who had set up a 9-inch telescope near a ramshackle cabin near the summit. It was the same cabin Milton Humason would find later that summer on his first trip to the mountain. Hussey had been commissioned by Carnegie to explore sites in the United States and abroad the year before and had decided that Wilson's Peak was as yet the most suitable location for a new observatory. Mount Palomar, further south near Oceanside, was by all indications even better but was deemed too remote by the careful and accomplished observer to be a viable option.

The two days he spent on Mount Wilson would be remembered by Hale throughout his life. The climate in southern California and the quality of the seeing on the mountain impressed him greatly. The views from its summit overlooking the Pacific made it all the more inviting, and the beautiful community of Pasadena would make a wonderful home for his family. All in all Mount Wilson was an ideal location for a new modern observatory. As he made his way back down the mountain, Hale began wrestling with the daunting challenge of convincing Andrew Carnegie of the soundness of his plan.

* * * * * * * * * * * * * * * * * **

By the spring of 1904, life was beginning to take on a routine for Milton and family. The year before they had moved into a one-story home at 1314 Kellam Avenue while they awaited the completion of their new home around the corner at 1345 Carroll Avenue (just west of modern-day Dodger Stadium). The two-story Victorian was located in the beautiful Echo Park area of Los Angeles. When finished the house would feature a large front porch framed in latticework, tall windows and a long stairway leading up from the street. The elevated front yard was buttressed by a 3-foot-high stone wall on the street. At that time Echo Park was as far north as civilization had spread; virtually everything around them north of the city was fields. The tree-lined park was a great place to spend an afternoon boating or cycling around the bike paths. A wooden footbridge led visitors to an inner island where a gazebo had been built. There they could spend the afternoon sipping tea and listening to music. The Witmers, who lived just a few miles south of them, were located near the equally serene Westlake Park (yes, there was an Eastlake), and the two families enjoyed afternoons in the park no matter which home they were visiting.

The family's fortunes had changed for the better since the move to California. William's health had improved significantly, and he was beginning to get acquainted with his brother-in-law's financial interests. That year, Henry Witmer had merged his interest in the bank to form the American National Bank with himself as vice-president. William's expertise in bookkeeping made him the easy choice to manage the books. A proud man, William took no special favors from his millionaire brother-in-law. A longtime member of the Chamber of Commerce, Henry was active in building projects all over the city, and William was grateful to have his opinion on the best areas of the city to house his family. The untimely deaths of his brother and brother-in-law left a void that Henry Witmer would never quite manage to fill, but he was glad to have William's support to the extent he could be useful. Having his beloved wife's older sister around was equally helpful, and the addition of the Humason's 5-year-old daughter, Virginia, gave their son, Joseph, who was the same age as Ginny, a sibling companion of his own. This companionship would continue into their teenage years and adult lives.

For Milton the long-awaited day finally arrived when he was to head back to summer camp on Mount Wilson. Life in America was brimming with hope and innovation. In November, President Theodore Roosevelt had pushed his plan for the creation of the Panama Canal through Congress, and work was already beginning on the ambitious project. On December 17, the long effort to record the first-ever manned flight of a heavier-than-air machine was accomplished by a pair of bicycle makers named Orville and Wilbur Wright near Kitty Hawk, North Carolina. One of the flights lasted a minute, covering a distance of nearly a quarter mile, and officially ushered in the Age of Aviation. The brothers would patent their flying machine design two years later.

Despite all of this innovation, all Milton could think about was fishing. He couldn't wait to get back to Mount Wilson for summer camp and a chance to uncover more of the mountain's secrets. He had learned on his visit the year before that the secrets the mountain held were many and magical, although often simple. One of these was a set of pine trees he had found just off the trail near the Echo Rock overlook. Mount San Antonio, known as Old Baldy, was set in the distance, and the exposed roots of the two trees made an excellent seating area from which Milton could sit and ponder the enormity of nature and the vastness of the sky above. He had spent many nights there doing just that since first discovering it the year before.

This summer, however, something vastly different was in the works on Mount Wilson. On a hike near the summit where the old log cabin stood Milton and his friends noticed there were men working on a strange new contraption. An array of mirrors stood at one end of a high canvas tent that stretched some 60 feet back from the edge of the hillside. The other end of the tent was attended by a man wearing a 20-gallon cowboy hat, high boots and a pistol with a full belt of ammunition. The man with the handlebar mustache who was peering out through wire-rimmed spectacles was Ferdinand Ellerman, and the contraption he was overseeing was the Snow solar telescope, imported from Illinois with permission from the University of Chicago. Nearby construction was underway for a new elevated pier to house the telescope.

Milton noticed that the roof of the old cabin had been sealed, and the area around it had been cleared of ground cover and debris. His Uncle Clate had mentioned that there had been some activity on the mountain earlier that spring. News had leaked in April about the arrival of George Hale on an expedition from the Yerkes Observatory to explore the summit. Further news that the expedition was being funded by the Carnegie Institution of Washington had caused a buzz around the city, and now it was stirring up excitement around Strain's Camp.

The arrival of Hale and his team from Yerkes Observatory marked a turning point in the history of Mount Wilson, but, for Hale, the outcome was still uncertain. He was a careful student of the funding game and knew he held the upper hand if Carnegie could be convinced to fund an entirely new facility. At the moment, the aging philanthropist was holding his cards close to his chest, as the institution that bore his name busied itself with the work of funding a vast array of scientific problems. Gambling on a positive outcome, Hale had immediately begun making plans to start operations in Pasadena after his first visit. He had moved his family to the area and, after being delayed due to weather, joined them in December. Hale arrived back in Pasadena having been awarded the Gold Medal of the Royal

Astronomical Society for the invention of the spectroheliograph and the increased investigations of the Sun the new instrument of science had produced. He had also been elected to the National Academy of Sciences, one of the youngest men ever elected to that distinguished council.

In another daring move, Hale had sent for the Snow telescope and was in the process of erecting a new elevated pier for it using his own money while he awaited word from the Carnegie Institution on funding. The horizontally designed telescope, it was discovered, must be set above the mountain floor to avoid surface heat that could corrupt the Sun's image on the plates being exposed. Hale discovered this early one morning on a trip up to the summit. He was escorted by a 14-year-old boy from his neighborhood in La Solana named Seward Simons. Carrying with them a 3-inch telescope Hale first measured the seeing at ground level but decided he needed to test this against the seeing at a higher elevation. The only thing handy was one of the yellow pine trees that made up much of the surrounding forest, so Hale scaled one of the trees, dragging the telescope up behind him. He took measurements at 32 and 68 feet while anchored into the limbs of the tree and discovered a marked improvement in the seeing at these elevations. These findings later informed his decision to elevate the Snow telescope as well as future solar telescopes at Mount Wilson.

Early in 1904, Hale had worked to secure control of other important aspects of the operation of the mountain. He negotiated land rights and the use of the mountain toll road for transporting materials and supplies up to the site. Ellerman had been the first of his team from Yerkes to arrive in Pasadena. Hale had sent for the jack of all trades to help him get started setting up the solar telescope and clearing the way for further developments. A superb observer with a colorful personality, Ellerman was a master at making repairs to instruments using whatever he had on hand in order to keep them operating. Later, Hale sent for his right hand, Walter Adams, and the superior telescope designer and optician, George Ritchey. The latter was employed in the building of two laboratories, one on the mountain and one in Pasadena.

Progress on the mountain was good, but indecision among the Carnegie board members and the sheer magnitude of the job at hand began to deplete Hale's finances by year's end. With money running out and his hopes fading Hale's dream finally came true. On December 20, Hale stopped at Martin's Camp on his way to the summit and was summoned to the single-wire telephone at the camp. The message from the other end of the line was that the board at Carnegie had granted $150,000 per year for two years and called for the immediate execution of the larger plan at Mount Wilson. Hale was elated. After the anxiety and the stress of the past two years he had finally succeeded, and the Mount Wilson Solar Observatory was officially founded.

Unlike its predecessor, the Lick, Hale's new observatory would be designed for limited visitation rather than year round living. Problems had arisen at the Lick, where staff and families lived and worked on the mountain. Inclement weather, the need to build stores, hospitals, and stores, and other perils of living on the mountain sapped precious resources needed for research. Hale's idea was to create the conditions for long but limited use by a revolving staff of observers on both solar and eventually stellar observing runs.

There was much to be done to prepare the mountain for the operation. At the summit, trees had to be cleared and large stones uprooted and transported for the creation of piers and foundations for buildings. Water from the nearby San Gabriel River was diverted to a reservoir built into the mountain to provide water for cooking and to fight fires. Breaks were cut along the face of the mountain to prevent forest fires leaping up the mountain and devouring the observatories' buildings and equipment. The long trail that ran for 9 miles to the summit from the base of the mountain had to be widened in order to increase the size and amount of traffic up and down hill. By the beginning of 1905 teams of mules strapped with long boards for building construction made their way up the steep mountain trail every day. The narrow winding road made it impossible for the mules to carry any board longer than 10 feet in length. The mules also hauled wagonloads of tools and hardware, steel, cement, stone and other supplies to the summit. Areas of the trail that were too narrow for the larger materials and equipment to pass had to blasted. Delays were frequent. Excessive rain caused frequent mudslides that covered the road, rendering it impassable. Winter brought periodic snowfall that buried the mountain in snow. Because the trail wasn't wide enough for two-way traffic, a phone line was strung from the base of the mountain to the summit to coordinate movement up and down hill.

* * * * * * * * * * * * * * * * * * **

In the summer of 1905, Milton Humason returned to Mount Wilson. For the past few years he had been spending his summers roaming the hillside, roughhousing with friends, plunging into the cool water at the swimming hole and having one adventure after another. It was a dream world for a growing boy, but Milton was starting to get pressure from his mother and father to focus more on his schoolwork. The improved condition of the family's finances might enable William and Laura to provide an education beyond high school for him and, perhaps, even an advanced degree. In any event, it was time for the oldest of the three children to start thinking about what his future would look like in the working world. He would be entering high school in the fall and would be expected to apply himself more thoroughly in the coming years.

The few years he had spent getting to know his Aunt Alice and the Witmer family in Los Angeles had made Milton forget all about his early childhood home in Winona. Laura and Alice kept in constant communication and the families met frequently for weekend outings on the Witmer family farm in Lordsburg. Milton loved visiting the farm, riding horses, picking oranges in the vast orchards, and hiking and fishing in the nearby hills with his brother Lewis and their cousin Ralph (Fig. 3.1).

For Milton, the pressure to heed his parents' wishes to continue with his schooling was proving no match for the allure of a life on Mount Wilson. Though his parents used every opportunity to impress upon their son the need to get a good

Fig. 3.1 Reading in the sitting room. Seated from *left* to *right* Joseph Witmer, Lewis Humason and Milton Humason. Virginia Humason sits on the floor in front of them

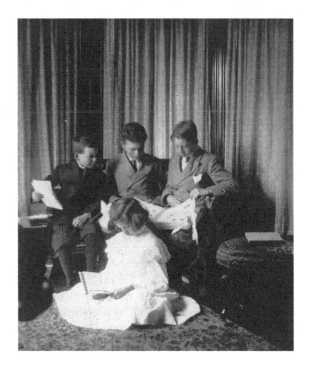

education, Milton was never an eager student and found he had little use for his studies; the more time he spent on Mount Wilson the farther he felt from the world of banking and real estate that had become the fabric of his family's life in Los Angeles. To the young and adventurous teenager, the richness of the natural world far outweighed these familiar trappings.

The fact that a modern observatory was being built at the summit of Mount Wilson made it almost too good to be true. Since before the onset of construction of the first buildings on the mountain, George Ellery Hale's observatory was the talk of the town. Hale's energy and enthusiasm lit up the pages of the local newspapers as he spoke of the great discoveries the new facility might have in store in the future. Public outreach was a great source of exposure and made for good press that smoothed the brows of Carnegie Institution board members, which furrowed whenever construction had to be halted due to a landslide, a snowstorm or a supply shortage.

Over the past several years the mountain had begun to change, and Milton had seen its development with his own eyes. The more he saw the more he wanted to be involved. Activity at the corral at the head of the trail was at a peak as muleskinners worked to secure the loads of building supplies and other materials needed for the construction of the observatory's first buildings to the backs of mules. Everything imaginable from pots and pans to chairs to women and children were hauled up the mountain on the backs of these lazy, often maligned but steady and, ultimately obedient beasts of burden. In those early days on Mount Wilson they were the sole means of transport. Some of the mules were loaded up with boards up to 10 feet long strapped to their sides while others were harnessed to small carts or wagons loaded with limestone and other supplies. Still others carried tools for digging and construction. The strongest mule was capable of transporting a load of about 200 pounds. The wagon, created specifically for passage up the narrow trail, was capable of holding 1000 pounds of material and was hauled by a team of mules in a line harness.

At the peak a newly completed Snow solar telescope rose from the hillside on its great stone pier. The instrument, which had been donated by a woman named Helen Snow in Chicago, featured a 12-inch coelostat mirror, mounted parallel to Earth's axis; it reflected the Sun's rays along a horizontal tube to a 6-inch photographic lens that formed the image on a plate some 62 feet away. When the Mount Wilson Solar Observatory was founded in 1904 the telescope was its feature attraction.

All around the summit there were signs of growing interest in Mount Wilson and its observatory. One of these signs was the creation of the Mount Wilson Hotel. The visitor's lodge had been built inside the old log cabin, the "casino," which had been newly renovated the year before and was now open to the public. The Mount Wilson Toll Road Company had bought the rights to the grounds surrounding the hotel and built small cabins around the lodge to host overnight visitors. A sign posted on the wall outside announced the need for bellman and other personnel to help operate the hotel. The rustic single-story cabin was made of large square-cut timber under a corrugated steel roof and was set back on a 3-foot-high earthen pier lined with stones. A small set of log steps led to the dirt porch that was lined with chairs where visitors could sit and talk or enjoy the outdoor air. The outer walls of the hotel were adorned with pelts and signs indicating upcoming events and attractions. To the right of the door to the hotel was a small canvas-covered counter that served as the hotel's office. Here visitors checked in and paid for their stay at the hotel, which featured several small rooms inside and a variety of shacks and camping areas for visitors to pitch tents nearby in the forest. Inside a mixture of comfortable leather and wicker chairs and rockers surrounded the hearth to a great stone fireplace at the center of the lobby. A table sat on the narrow planked hardwood floor and candelabras hung from the ceiling at either end of the room. On the wall at the back sat an upright piano where players of varying talents could entertain themselves to the relative delight or dismay of their fellow guests. Dances and special events such as Indian Day were frequent occasions on Mountain Wilson. The hotel lobby was cleared for dancing and musicians brought into provide the entertainment on these occasions (Fig. 3.2).

By now, the name George Ellery Hale was well-known both in Los Angeles and on Mount Wilson. Although slight of stature, the 37-year-old observatory director had a steady, penetrating gaze, a full mustache and energy to match the Sun. He could be seen at the trailhead in the early mornings making sure the details of the day were in order before setting off up the trail on foot. In the early days on Mount Wilson Hale never rode up the mountain, preferring, instead, to hike the 10-mile trail with colleagues and friends, all the while holding forth on subjects ranging from the arts and science to history and politics. A fan of the poets Keats and Shelley, and composers such as Bach and Beethoven, Hale would recite poems or

Fig. 3.2 The Mount Wilson Hotel shortly after it opened in 1904

hum melodies from his favorite symphonies. Hale was, himself, a captivating and invigorating force of nature who, through sheer will and determination and with a strong sense of his own convictions and deft understanding of the rules of negotiation, was rebuilding Mount Wilson in his own image. He was the driving force behind everything happening at the observatory, which was invested in public outreach and led regular public viewings of the stars and tours of the observatory grounds with visitors to the hotel.

Young Milton Humason was one of those captivated by the observatory and its charismatic director. By the first day of his freshman year of high school Milton had all but made up his mind that he wanted to be a part of the excitement and growth on Mount Wilson. Los Angeles High School, on Hill Street between Bellevue and California streets, was just a mile or so from home in Echo Park. The young school was boiling over with activities at the start of the new year. Tryouts for sports were listed, book and chess clubs were having their first meetings, the marching band was having auditions and class leadership elections were being planned. One club of note to the young explorer was the Kodak and bicycle club, which roamed the city and surrounding areas on bikes, stopping at intervals for photographs of areas around town and country. The Eastman Kodak Company introduced the first consumer brand camera, the $1 Brownie, in 1900, opening the emerging field of photography to the masses. Seizing on the opportunity, the school had created the club to provide a new source of extracurricular activity to interested members of its student body.

As he wandered the halls of his new school that day, Milton Humason's mind was elsewhere. Perhaps it was the change in settings at his new school, or maybe it was just simply the call of Mount Wilson, that wondrous and familiar forest wilderness he had grown to love, that gave him the courage to do what he did next. With the mountain standing in the distance over his right shoulder, Milton rode his bike home from school that day and sat down with his parents to plead his case for leaving school to take a job on Mount Wilson. The hotel on the mountain was hiring. He would stay on the mountain during his work week at the hotel and ride his bike to the trailhead and then stay on the grounds of the hotel during each work run. At the end of each run he would hike back down the mountain and ride back to spend his downtime at home. His parents weren't keen on the idea of letting him quit school, but after some persuading it was decided that they would try the arrangement for a year, and if he didn't like it he would go back to school and dedicate himself to his studies. His first day of high school turned out to be Milton Humason's last day of formal education.

* * * * * * * * * * * * * * * * * **

George Hale had his hands full with the details of getting his newly formed observatory off the ground. In an essay to the Carnegie Institution board in 1903, Hale had written of the prospects for understanding stellar evolution if a solar observatory, utilizing both day and night observing time and the best available equipment, could be built. Now a year into his directorship of the Mount Wilson Solar Observatory, Hale and his chief assistant, Walter Adams, were busy with the work of preparing for the arrival of the 60-inch reflector.

Although the Sun's location so close to Earth made it a favorite target, Hale knew that a thorough understanding of stellar evolution would require much study of stars outside the Solar System. Unlike the world's other big reflecting telescopes, such as Lord Rosse's 72-inch reflector (nicknamed the Leviathan of Parsontown) at Birr Castle in the county Offaly, Ireland, the 60-inch would be able to track stars at all points of the night sky. The advantages of reflecting telescopes had been known to Hale for some time. George Ritchey's 24-inch reflector at Yerkes had confirmed the superior light-gathering capabilities of reflecting telescopes. With the Yerkes 24-inch reflector direct photographs of stars, nebulae and other objects well beyond the reach of the observatory's big refractor were made possible. If its design and creation were successful, Hale would have created a telescope with six times the light-gathering capabilities of the 24-inch reflector at Williams Bay.

This was much easier said than done. The crews that worked in the initial phases of the operation of the observatory had been having enough trouble transporting the materials for the solar telescope and smaller buildings up the steep and narrow trail. When complete, the telescope and dome would weigh some 150 tons. No mule team could haul the heavier steel beams up the mountain. Even the 1000-pound mirror would be a problem for the muleskinners and their teams. A stronger vehicle was needed to transport the heavier components of the telescope and dome to the summit, and the toll road would have to be widened to allow for the vehicle's size and weight. The mounting for the telescope was being cast at the Union Iron Works in San Francisco while, in the observatory's new lab on Santa Barbara Street in Pasadena, George Ritchey busied himself with the exacting work of grinding the giant glass disc for the 60-inch mirror. On the mountain, a location for the new telescope was selected on a high slope south of the solar telescope and, the work of first clearing and then excavating the area, to make way for the new dome, was underway.

Protecting the delicate instruments and equipment at the observatory against forest fires was paramount. Fire breaks were created with the assistance of the local forest preserve and a new 30,000-gallon water reservoir was built near Strain's camp with a fire protection system attached. The reservoir's main purpose was to provide water to the existing buildings.

A new laboratory, 27 × 70 feet, was being built with furnaces and other instruments for the study of the radiation of gases related to astrophysical research. A nearby warehouse was used to store tools and equipment used in the creation and maintenance of the facility's growing number of buildings.

Down the path from the observatory stood the new dormitory, nicknamed the "monastery," where staff ate and slept during their stay on the mountain. The dorm had electricity and a phone line connecting observers to the new offices in Pasadena. In addition to the dormitory a guest house built by John D. Hooker, a hardware magnate from Los Angeles who Hale had met through his connection with Campbell at the Lick, was used for out of town visitors during their stay on Mount Wilson. The two-bedroom house featured splendid views of the valley from its wide balcony.

The shop and powerhouse were built to house the gasoline engine, dynamo and switchboard used to power the various machines and equipment. A bank of thirty batteries stored available electrical current for use around the facility. A second smaller power plant had been set underground in a 13 × 13 room to help keep its attendant array of batteries cool during the hottest days of the year. The work in the shop consisted of maintenance on the instruments and other equipment as well as the creation of tools for the ongoing construction of the observatory.

* * * * * * * * * * * * * * * * * **

The rain would probably not stop for some time. Milton put down his book and leaned over toward the fire crackling in the fireplace inside the hotel lobby. The heavy rains could cause landslides that rushed down the mountainside without a moment's notice, sweeping away everything in their path. For that reason Milton knew there would be no foot traffic on a day like today. A slow day was welcome once in a while, and the damper weather would be ending soon now that April was approaching. The steady rainfall did mean it was likely they would find some new leaks in the hotel's roof. If found, these would require mending once the weather cleared.

In the six months since he had been hired as a bellboy at the Mount Wilson Hotel, Milton had been making his way up the mountain every week for work.

Young and amiable with a quiet confidence and quick wit he applied himself ably to whatever he could to contribute to the buzz and excitement that seemed to build constantly at the observatory. Mending fences and other work he had done around the house and family farm with his father and uncle had turned the youngster into a competent handyman, and Milton was more than happy to help out whenever possible around the hotel. For a motivated soul, a mountaintop hotel and its grounds always offered one task or another to accomplish. Milton painted walls and fixed doors and patched holes in the roof of the hotel and other small cottages on the grounds. But happy as he was to have the work, his mind was on another prize.

The close association between the hotel and the observatory meant that he was exposed to many of the issues and developments surrounding the facility and the characters that inhabited it both near and far—men such as Ferdinand Ellerman, the observatory's fix-it man, solar observer and photographer. A gifted tinkerer with an imaginative spirit, Ellerman was, after Hale, the most charismatic of the original five members of the observatory, which also included Walter Adams, George Ritchey and Francis Pease. Although they had living quarters at the dormitory on the observatory grounds, their friends and relatives often stayed at the hotel during visits to the summit, enjoying afternoon walks through the forest or viewing the images being produced of the Sun and stars with the telescopes. Nearly every day visitors to the area would ride into the compound on the backs of burros or stride in with their backpacks over their shoulders seeking a night's stay at the hotel while they explored the summit and surrounding wilderness. They were always escorted by a cowboy on horseback who would collect his stock once dismounted and return them to the nearby corral, where they were fed and watered after a hard day's work. Train after train of mules would steer its way into camp hauling or towing the seemingly endless supply of materials being used to build the observatory buildings. Still more trains brought provisions for the hotel, all of which Milton and his fellow clerks at the hotel offloaded and stored for later use.

Work at the hotel was fine, but Milton had his heart set on becoming a muleskinner like the young men who road in and out of the hotel receiving yard every day. He would be 15 years old in August, and he could ride a horse as well as any of the cowboys he saw riding in and out of camp. It was just a matter of time before he joined their ranks.

On his return home to Los Angeles Milton discovered his Aunt Alice and Uncle Clate had returned home with young Joseph from their recent trip. After a train ride across the southern end of the country the family had sailed to Cuba, the tiny U.S. protectorate in the Caribbean Sea. After spending some time exploring the island nation they had boarded the S.S. *Prince Arthur* and sailed to the port of New Orleans, landing there on February 19, 1906. There they stayed to see the sites of the town and to watch the yearly Mardi Gras parades and celebration before loading themselves onto a train for their return home. The trip was planned in part to celebrate Uncle Clate's fiftieth birthday and probably the Witmers enjoyed taking 7-year-old Joseph on his first ocean voyage.

The trip to the Caribbean was just what the family needed to restore itself after the trials of the past few years. In the wake of his brother's deaths H.C. Witmer had

Fig. 3.3 Families united, the
Humasons and Witmers stand
on the porch of the Witmer
mansion on 3rd Street in Los
Angeles

spent all his time and energy keeping the family's varying business interests active
while balancing home life with his wife and young son, not to mention his sister
Mary and her family home. Although he never complained Alice knew the burden
was too much for her husband to take on alone and pleaded with him to slow down.
William's help around the house and business was much appreciated, but there was
only so much he could take on with a wife and three children of his own. In general,
the mood of the family was good and spirits were high (Figs. 3.3 and 3.4).

At about 5:12 a.m. on Wednesday, April 18, Milton felt a long tremor that shook
the ground beneath him inside his tent on the hotel grounds. He stared into the
darkness at the tent liner, waiting for an another tremor to rumble through. When
none came he shut his eyes and went back to sleep. If you lived in Los Angeles long
enough you were going to have to get used to the fact that the earth there periodically shook. Later that morning, as the Sun rose over the mountain, news began
to circulate of a major earthquake in San Francisco. Stories from inside the disaster
were spotty at that point because many of the city's phone and telegraph lines were
down. For several long days Milton and his family waited for word from their loved
ones who were living in the area. As was learned later, the earthquake had ruptured
the gas mains that coursed the city's interior, starting a number of fires that soon
engulfed the city. With the water lines also broken, the city's fire departments were
unable to extinguish the fires using conventional means and had to resort to creating
fire breaks. Dynamite was used in many cases, which caused still more fires. In the

Fig. 3.4 Clowning around at the ranch (1904). Milton sits on a tree swing with Lewis, Joseph Witmer and Virginia on his shoulders, while William Humason watches attentively

end the earthquake and its resulting fire destroyed 80 % of the city, killed thousands of people, and displaced about three-quarters of San Francisco's population of 400,000 people. The Humasons and the Pettersons survived the catastrophe without any loss of life.

* * * * * * * * * * * * * * * * * * **

George Hale was desperate for news from the Union Iron Works in San Francisco. The largest steel manufacturer on the West Coast, the iron works was part of the Bethlehem Shipbuilding Corporation, and its shipyard was located across the bay. Hale hoped that this fact had somehow saved it and more importantly his new telescope mounting from destruction. Walter Adams had sent word from Pasadena to Hale, who was on the East Coast the day of the quake. Upon reading the news, the observatory director rushed back to California only to find there was no news as yet from the fire-ravaged city on the bay.

Finally, a telegram came later that afternoon. The earthquake had torn two warships, in the process of completion at the iron works, from their stocks. Another ship, the cruise liner S.S. *Columbia,* which had been brought to the hydraulic dry dock for repairs, had rolled to her starboard side during the earthquake, causing irreparable damage to the dock. The ship had taken on water, and crews were desperately working to get the foundering ship righted again. Fortunately for Hale and the observatory, unlike most of the rest of San Francisco, the manufacturer had been spared the fire that had torn through town and destroyed everything in its path. The mounting for the 60-inch telescope was still intact. Unfortunately, it would be some months before the factory dug itself out from the earthquake and could begin building again.

Hale was relieved at finding out the telescope had survived the disaster and that no lives had been lost, at the iron works, anyway. Although it had been a work day, the early hour meant that no one was at the factory when the earthquake struck. On such a grim occasion any news was good news. In another stroke of luck Hale and Adams heard from Campbell at the Lick. The observatory had been spared completely, and all members living and working on the mountain were busily about their work. In the days following the earthquake and fire, reports of the events that unfolded as the fire gripped the city started to come in. Surreal scenes of people watching the smoke and flames from nearby hillsides, of frightened townspeople fleeing their homes for sanctuary by the water's edge and the heroism of those who had fought for and either survived or lost their lives trying to save their loved ones, were numerous. It was the kind of event so horrific that it defied the imagination to comprehend.

As the staff at Mount Wilson got back to business, it was decided that the weather was clearing, and the ground pack was getting firm enough, to resume the task of widening the road along the "old trail" to the summit. The road construction was being supervised by Godfrey Sykes, on loan to the observatory from the Department of Botanical Research and to this point had been about half completed. A crew of 120 men working in three shifts worked daily to widen the road, blasting away at the granite face of the mountain and clearing trees, dirt and flora wherever necessary.

It was hard to imagine, but all of the incredible development at the observatory was superseded that year by an even more singular event. Ever since their first meeting, Hale had noticed John D. Hooker's interest in the events on Mount Wilson, and he was keen to know what news Hale had from the mountain whenever they met. Hale had also made friends with Hooker's young wife, who was a talented writer and a gifted conversationalist in her own right. During one of the early meetings between the two, Hale had explained his philosophy on the course of discovery in the field of astrophysics, the lessons of mentors such as William Huggins, and his singular interest in the pursuit of creating instruments with ever more light-gathering capability. In time, Hale envisioned, that these instruments, in the hands of the right people, would unravel the secrets of the universe. Hooker was impressed by the poetic tone, intelligence and sheer passion of his counterpart and decided to grant $45,000 to Hale for the creation of a mirror of a 100-inch aperture. The mirror, Hale would later say, would have a focal length of 50 feet and would be housed in a great dome on the mountain. Once set in a suitable mounting, the great telescope would have 2.7 times the light-gathering capability of its smaller and as yet unfinished cousin.

George Hale was not one to sit on his hands. He immediately sent word to the French Plate Glass Company in St. Gobain, France, informing them of his intention to create an enormous new glass disc that could then be ground and polished to form the mirror blank. Upon the completion of the 60-inch on Mount Wilson, the observatory's design team would use the experience of operating the new telescope to inform them as to the design of its larger cousin.

* * * * * * * * * * * * * * * * * **

A massive snowstorm on January 10, 1907, dumped over 5 feet of snow on the upper half of Mount Wilson, shutting down the mountain, causing landslides, and downing electrical and phone lines. The storm was preceded by days of rain and snow rarely seen in the area. Milton was at the family home that day in Echo Park. During the summer of the previous year Milton had informed his parents of his desire to keep working at the hotel on Mount Wilson and forego any further formal education. The work at the hotel was steady, and he intended to join the ranks of the muleskinners as soon as he could. Compulsory school laws enacted in most states by 1900 required children to be educated until the age of fourteen, but education was still not regarded as necessary in the early years of the twentieth century. Often the necessity of keeping a roof over the family's head and providing basic needs led to children being sent into the workforce at an early age. Family tragedies, such as the death of a father or mother or both, were sometimes the incentive for these decisions. In the case of Milton Humason, who would have perhaps had certain advantages others would not due to his family ties, the decision was made out of sheer passion for his mountain home away from home.

The extent of new construction at the observatory had led to an increased demand for workers of all types. During stints on the mountain Milton had made friends with many of the workers and was earning a good reputation among the group working the mule teams on the trail. Several of them suggested that if he was interested in joining their ranks that he ask the stable manager, Mr. Bassett, for a job. That year, on the eve of his sixteenth birthday, Milton Humason applied for and was given a job as a muleskinner on Mount Wilson. In addition to the work at the stable, Milton would keep his job with the hotel as well, which gave him free room and board on the days he worked there. This would enable him to spend even more time on the mountain in the future.

Wearing his Mounty hat, a loosely fastened tie tucked into his long-sleeved button-down shirt, with gaiters at the bottom of his trousers for protection against mud and rattlesnakes, Milton beamed with excitement at his new job. He spent his early days walking the trail with his mules in tow, cajoling them to move along when they got lazy. It was no secret that he loved the work, and he was more than willing to help out whenever trouble arose. The best part of any day was when the mountain was visited by Hale and his associates, most of whom were very generous and respectful of the muleskinners and their charges. Hale would often ride his bike to the trailhead and hastily but with a smile on his face summon a burro to carry his things to the summit. Occasionally he would ride, but he preferred to hike whenever he felt up to it, which was most days. As the months progressed Milton learned how to care for the mules, their temperament and personalities. The more time he spent with them the more he earned their respect and the respect of the older cowboys on the trail. Milton began to make personal notes of the visitors to Mount Wilson in those early years. He regarded the mountain as his home, and as such, he didn't appreciate people who mistreated it or were otherwise disrespectful him and his fellow teamsters. If a fellow rubbed him the wrong way he made sure that some part of his journey to the summit was unpleasant. This was easy enough to do when you knew the animals you were dealing with and their behaviors. If someone rubbed you the wrong way you could accidentally forget that the burro you were giving him liked to puff out its chest while being saddled so that the belt holding the saddle on relaxed when he relaxed again. The site of the burro running off with the infiltrator slung over to one side holding on for dear life could be quite amusing, and of course, Milton and his friends were always there to save the day. During these early days on the mule teams Milton began to hone an innate quality that would serve him well in times of crisis throughout his adult life, an ability to quickly measure a person's character. He also learned how to pull a prank on an unsuspecting victim without being caught. His slippery ability was due in large part to the fact that he very often acted on his own accord (Fig. 3.5).

Fig. 3.5 The young muleskinner, Milton Humason, stands for a photo with two of his charges

* * * * * * * * * * * * * * * * * **

On the surface there was really nothing to be anxious about for George Hale. But under the surface he could feel events were gradually starting to spin out of control. A year earlier he had been asked to assume the office of the president at his alma mater, M.I.T., but turned it down to pursue his life as a researcher and observatory builder. He didn't let on that he was having more and more sleepless nights and suffered from anxiety on a regular basis. Still wanting to be involved in continuing education Hale had recently joined the Board of Trustees at Throop Polytechnic Institute in Pasadena. Hale envisioned that someday the school would become the M.I.T. of the West Coast.

There was much to cheer about at the observatory as well. Benefactors seemed to be lining up to give him money for his many scientific schemes, and he had started not one but two projects to build the largest sidereal observing instruments in the world in close succession. The 60-inch was nearly completely assembled in the assembling house in Pasadena, and the road to the top of Mount Wilson had been completed in May by his superintendent of construction, George Jones, a giant of a man who could toss boulders the size of Hale himself with his bare hands. But unforeseen delays caused by the earthquake in San Francisco the year before, striking laborers at the Union Iron Works and the snowstorm of the past January had backed up the delivery of the new telescope to the mountain until 1908, and added significant cost. These cost overruns had to be explained to the board at Carnegie.

The progress in the spring on the new reflector and the road to the summit gave Hale a chance to head to France to visit the French Plate Glass Company. In a development all too familiar to Hale, the millionaire behind his latest brainchild, the 100-inch telescope, was growing impatient for word on the creation of the glass disc that would be used for its mirror.

Hale arrived in June and met with the director of the company, Mr. Delloye, on the design and plan for casting and annealing the 4.5-ton disc. He learned to his satisfaction that the first attempt to cast the disc would take place in July and, although the annealing process would take some months in order to prevent strain on the glass from cooling too quickly, there was a good chance they would have a resulting disc before the end of the year. Whether the result would be positive or negative remained to be seen.

Back in Pasadena progress remained fast and furious during the summer months. Having completed the road to the top of the mountain, George Jones and his crew were finishing work on the concrete foundation for the new telescope dome and mounting pier. Power and phone lines damaged in the snowstorm had been buried under ground for miles along the roadway and a storehouse had been built near the monastery.

In one of his favorite developments, the new 60-foot tower for the solar telescope was just being completed near the Snow telescope building. Shortly after the Snow telescope came online in the summer of 1905 Hale and his five deputies, Adams, Ritchey, Ellerman and Pease, whom Hale had successfully plucked from the Yerkes Observatory the year before, decided that turbulence caused by heat at the surface of the mountain was inhibiting the telescope's ability to capture steady images of the Sun. Building high, they believed, would allow the turbulent surface air to be swept away by the upslope breezes from the Pacific Ocean. In the wake of this decision

Hale had designed the new 65-foot high steel-framed tower that was standing above the tree line. The coelostat mirror at the top of the dome reflected sunlight off a second mirror down through a 12-inch objective with a focal length of 60 feet There the Sun's image was formed near ground level, where it entered the slit of a 30-foot grating spectrograph that descended into an 8 1/2 foot wide pit deep in the ground. The vertical height and orientation as well as the natural cooling the pit provided were novelties that Hale liked boasting about whenever the opportunity arose.

The operation had revealed some unrest among the ranks of his staff, however. Francis Pease, the mild-mannered designer and observer who designed the new solar telescope had resigned his position not long after he'd finished it when his request for a pay increase was denied. Walter Adams, also a man of level head and mild demeanor, didn't see eye to eye with George Ritchey, who was brilliant but stubborn, obstinate and overbearing. Hale had his own run-ins with his laboratory superintendent over the issue of the course of research at Yerkes and Mount Wilson. In the end, the savvy director had always taken the congenial approach, allowing time for nerves to be calmed and the order of business (Hale's order of business) to be resumed. For now, Hale had no reason to counteract that course of action.

The other thorn in Hale's side was the massive coupled-gear truck the observatory had purchased to haul the heavy steel pieces for the new telescopes up the mountain. The truck had arrived in November of 1906. It was essentially a flatbed on a steel frame with an extremely short wheelbase to allow it to fit around the tight turns on the mountain trail. Each of its steel wheels was powered by an electric motor, and these were, in turn, supplied by a 17-kW generator (refitted from a 12-kW generator) that was charged by a 45 horsepower Brennan engine (refitted from a 25 hp engine). After a year of testing and retooling and waiting for conditions on Mount Wilson to improve enough to test it again the truck was set against a mule team of four mules to see which could more ably haul a load of 2 tons to the summit. In southern mountains equivalent to Paul Bunyan in the north woods, the mule teams beat the iron wagon soundly. A four-mule team driven by one of the better muleskinners was far more efficient bringing this size load to the top of the mountain. The mule teams would from then on be employed in the hauling of such loads while the truck would be relegated to hauling only the largest loads up the mountain, which were too heavy for the mules to pull.

* * * * * * * * * * * * * * * * * **

On a sunny afternoon in 1907 a shiny new Franklin automobile climbed the Mount Wilson Toll Road and drove onto the grounds of the Mount Wilson Solar Observatory. The occasion marked the first time an automobile had made it to the top of Mount Wilson. Now, not even a year later, the Ford Motor Company was mass-producing their new auto, the Model T, on its innovative new assembly line at its factory in Detroit. On family visits Uncle Clate had spoken enthusiastically about the future of automobile transportation. His position and contacts in the business community gave him access to information not always available to the layperson. Himself an innovator at heart, Henry Witmer always made sure the family home in Crown Hill carried the latest and best new gadgets money could buy. By 1908 Aunt Alice was showing off her new washing machine.

However, the advantages of the automobile weren't quite as clear in the early days of its development. Milton and his fellow teamsters were skeptical that the arrival of the car on Mount Wilson, loud, stinking of gas fumes and spewing smoke, spelled the end for their way of life. Such a device might do well for day trips to the park, but it could never take the place of a well-managed mule team on the dusty trail to the summit. They had learned this first hand during the year when the new truck was being tested against their mule teams to determine its efficiency and viability for use in hauling the heavier pieces for the 60-inch telescope up the mountain. The personality and hauling ability of their charges was a point of pride for the muleskinners, and they would take every opportunity to remind those who would listen that they had won their battle against modernization.

The families were at the Witmer ranch, Aljoleto, for the weekend to celebrate Milton's seventeenth birthday. An afternoon spent riding the trails around the ranch, hiking with his brother and cousin and cooking out on a warm August day were a good way to spend a birthday before heading back to the mountain.

The 60-inch dome was nearly complete, and the new telescope was being assembled on its pier inside. As he approached the station house for the Mount Wilson Toll Road Company at the base of the mountain, Milton could see the big truck standing in the loading yard. The large skeletal steel tube for the telescope, 18 feet long and 6 1/2 feet in diameter, was heavily strapped to the flatbed. He had been working the mule teams on the mountain long enough by now to understand that getting this load to the top was going to be trouble. It was a good thing they were attempting it before the rainy season started.

Milton helped his fellow muleskinners gather a team of strong mules to help the truck tow the telescope tube up the mountain. A long chain was attached between the mule team's yoke and the front of the truck, which consisted of the flatbed and a steering wheel at one end. Driving and steering the truck, whose front and rear axles both steered to help the behemoth get around the tight switchbacks along the trail, was hard enough. Keeping the engine, generator and electric wheel motors operating at optimum capacity was an even more difficult task.

To help keep the truck in good running order an electrician had been enlisted from Gaylord, Blick and Vore Electrical Supplies in Pasadena. The firm had sent over one of its most gifted electricians, Merritt C. Dowd, a year earlier to aid in wiring some of the instruments and buildings, and to help operate the new truck.

Dowd sat behind the steering wheel at the front end of the flatbed truck as the muleskinners finished securing the mule team to the front end. When everything was in order the all-clear was sounded, and Dowd fired up the truck's engine. When he engaged the drive gear the truck lurched forward, and the mule team let out in front of the truck. The first mile or so was relatively easy, as the truck made its way more or less under its own power through the lower section of the trail. As they entered the steeper climbs and twisting turns of the mountain nearer the top the going got much tougher. The trucks extreme weight could cause the road to collapse at any time so Dowd paid careful attention to staying as close to the mountain as possible. A few brave souls actually rode with the truck standing on the back of the flatbed while watching for signs of trouble on the road ahead. There were several places where the newly widened road was still too narrow, and the truck occasionally had to be helped out as the mountain began to fall away beneath the outer wheels. Both the trucks front and rear wheels turned, making maneuvering around outside turns easier, but where they had to pass close to the mountain on an inside turn, which was often, the height and size of the load was so large that progress was halted in order to clear tree limbs and to dig or blast out sections of the mountain so the truck and mule team could pass. The arduous task of clearing the roadway necessarily slowed the journey, and the mule teams had to be taken out of the yokes and rested. The men on the crew camped out at Strain's Camp overnight and were stirring before dawn to get an early start on the day (Fig. 3.6).

After what seemed like a week the tube was finally delivered safely to its new home in the 60-inch dome on the observatory grounds. By the end of 1908 the grinding and polishing of the glass disc for the mirror was completed by Ritchey and his team at the lab in Pasadena and ready for its trip to the summit of Mount Wilson. Upon completion, the new mirror weighed about one ton and was polished with such precision that no part of its surface had a variance of more than 2 millionths of an inch. The question now was how to get the giant mirror, which could be marred by the touch of a baby's hand, up the rugged and winding face of the

Fig. 3.6 Tricky footing as the heavily laden flatbed truck makes its way up the narrow trail on Mt. Wilson toward the observatory, 1909

mountain. To ensure a smoother ride the mirror was swaddled in a padded box and frame, and tied securely to the flatbed of the truck for its ascent. The mirror's narrow profile made the trip notably easier than that of the telescope tube, and before long the mirror and housing were being assembled inside the dome on the mountain.

As the year ended Milton was already making a name for himself among the other cowboys on the mountain. Although he was younger than most of them, his attention to detail, problem-solving ability and work ethic quickly caught the attention of the burro stable manager, Chester Huston. Milton Humason was quickly growing from a boy into a young man, and, although he held onto his job at the hotel, his passion was working as a muleskinner on the trail. For Milton, it was a dream come true to work at the same corral that had so captivated him ever since his first visit to Mount Wilson. So passionate was he about becoming a muleskinner, in fact, he probably lied about his age to get the job. He is listed on two different census reports in 1910. On one of them, taken in April of that year, he lists his occupation as "packer burro train" and his age as 20 years old. Milton's parents also included him in their census report that year, citing his correct age as 18 years, as he would be four months shy of his nineteenth birthday in April of that year.

Whatever the circumstances of his hiring, Milton was becoming one of the most respected of all the muleskinners on Mount Wilson. His encyclopedic knowledge of the mountain and expert riding ability combined with a quick wit, charm and an attention to details was beginning to make him a favorite with the other cowboys as well as the staff of the observatory. He was already an accomplished horseman, having spent hours riding around the Witmer family farm in Lordsburg, and his friends within the ranks of the cowboys on Mount Wilson made finding someone to vouch for his age relatively easy. As the new year began Milt, as he was becoming known to his friends, was starting to think about making the mountain a more permanent home.

* * * * * * * * * * * * * * * * * * **

In July of 1908, George Hale had written to his friend and mentor, William Huggins, that there seemed little doubt of the existence of strong magnetic fields in sunspots. While Ferdinand Ellerman was away on vacation, Hale had spent several weeks that June studying the dark blemishes in the Sun using the 60-foot tower telescope. "I have been carried away by the solar whirlwinds," he reported to John Brashear, the astronomer and former director of the Allegheny Observatory in Pittsburgh, Pennsylvania, who had years before invented an improved silvering technique that was being used in mirroring modern telescopes of the age. In his note to the beloved elder statesman of the science world, Hale spoke of great solar prominences being drawn into these highly magnetically charged spots in a matter of minutes. Hale was elated and immediately began the work of confirming his discovery while putting a program in place for more thorough study of the Sun. To do this he would need an even more effective instrument.

The 60-foot tower telescope had solved the problem of surface heat turbulence but created a new one. On the more breezy days, wind above the tree line buffeted the tower frame and dome that housed the coelostat mirror, which blurred the image on the plates at the grating located in a shed at the bottom of the telescope. This meant that good viewing was only possible on less windy days, which isn't an altogether frequent occurrence on a mountaintop observatory. A new tower had to be developed, this time slightly higher than the original. To design the tower, Hale enlisted the help of his old family friend, the famous architect, D.H. Burnham. In a brilliant display of ingenuity, Hale and his team designed the 150-foot tower to have two separate frames, each mounted on its own foundation near the base of the 60-foot tower. The inner frame would hold the telescope mounting platform and mirrors at a height of 160 feet while safely enclosed within the hollow outer frame of the second tower, which held the dome and elevator that carried the observers to the mirror platform. This configuration allowed the outer frame to absorb the vibration caused by the wind buffeting against it leaving the inner frame and mirrors motionless and thereby increasing the number of good viewing days significantly. A 75-foot spectrograph and spectroheliograph dug in a 10-foot diameter concrete well deep into the ground would be used to capture the Sun's image. Digging of the well was set to begin after the spring rains subsided.

The grandness and scale of the new observatory left visitors to the facility in awe of the engineering prowess and sheer magnitude of Hale's vision. In addition to the solar telescopes the 5-foot reflector was now constructed and in the final stages of completion. The enormous new telescope and dome were staged prominently on the summit, and the gleaming rounded surface of the dome could be seen from Los Angeles in the valley. Plans for the creation of still larger, more powerful and more effective instruments were causing even greater commotion in the local and national community.

All of which should have been to the delight of Hale, who was taking ever greater steps in his lifelong pursuit of the study of the Sun and stars. But the stress it was putting on him was beginning to take its toll. The first casting of the 100-inch disc, which had been sent to Pasadena earlier in the year, was rejected by Ritchey as being unusable due to defects in its surface, and Hale had sent his instrument construction supervisor to Paris to help devise a plan for a more successful casting of the 4.5-ton glass blank. When he heard the news that the mirror blank was dysfunctional, John D. Hooker was incensed and began taking steps to disengage from the project. Hale had gone to great lengths to convince the aging hardware magnate that his namesake telescope would one day soon go into operation and that it would be the most technologically advanced instrument ever created for the observation of the stars. His reassurances had succeeded for the moment in assuaging Hooker's anxiety, but Hale was worried about the ultimate success of the endeavor. He knew the plan for the new telescope was stretching the limits of modern technology, and he had seen the difficulties inherent in casting large glass discs during his attempts to create a refracting lens larger than 40 inches back in Williams Bay. All of this was wearying the director, and he decided, at the urging of his wife and doctor, to take a leave of absence during the spring of 1909 and tour Europe. During this time he would visit the glass works in

Paris to see firsthand the plan that was being put in place for the casting of the glass disc for the Hooker Telescope.

* * * * * * * * * * * * * * * * * * **

As the Humasons rang in the New Year in 1909 William and Laura were planning a trip to celebrate their twentieth wedding anniversary. The trip from San Francisco to Hong Kong and Honolulu would take as long as three months, the standard time allowance for an overseas voyage of the day. William and Laura were looking forward to their trip abroad. With William pitching in at the office and William Howard Taft, another Republican, elected president, business was good and the future seemed bright. So when Henry Clayton Witmer died of heart failure at the age of 53, while working on the family's farm in Lordsburg in March, a hush fell over the entire clan.

Witmer's death resonated throughout the city of Los Angeles, as the city mourned the loss of the respected banker, realtor, merchant, newspaper publisher, land owner and citrus farm operator. To many Henry Witmer was a great civic leader who had led the campaign to widen Broadway and whose inventiveness and drive had helped to develop Los Angeles into a modern and progressive city. Henry Witmer was buried at Evergreen Cemetery in Los Angeles a few days after his death on March 3 as family and friends from around the community looked on.

For Milt and rest of the family the loss at home was felt most acutely. Aunt Alice and young Joseph were left in their large home without a husband and father to care for them. To make matters worse, the Witmers had been left with no one to run the family business (Fig. 3.7).

Immediately after Clate's death, William put the Humason home up for sale, and the family moved back into the mansion on Crown Hill so that Laura could care for her grieving sister and her son. William took on more of the work running the business, and Henry Witmer's older sister, Anna Victoria, also assumed some responsibility. At 55, Anna was two years older than Clate, and it was thought that she was next in line to take over the reins of the Witmer Brothers Company. William Humason took over as head of the house on 3rd Street and was made secretary at Witmer Brothers. Letha Lewis, daughter of Henry Witmer's oldest sister, Mary Agnes, assumed oversight of the family ranch in Lordsburg. In the aftermath of H.C. Witmer's death, Joseph Witmer's widow, Josephine, made a visit to the family with her daughter, Mary, and sons, William and David.

On a warm spring afternoon all three families got together for a cookout at Aljoleto, the family ranch. They assembled around a rustic wooden table near a large tent inside a stand of trees in the forest. Two wood-burning stoves nearby were used for cooking, and later the families posed for a photo. In it Letha and Ralph Lewis stand near the stove while Laura and Alice stand at the table. Joseph Witmer, 10, sits near his cousin, Ginny, who is standing near the tent. Milt stands at the back of the picture leaning against a tree with his father, who is sitting nearby with his brother and cousins, William and David. The sons of Joseph and Josephine Witmer would eventually assume the mantle of responsibility for the family legacy.

Fig. 3.7 Milt dressed in a
suit standing outside his
aunt's house on 3rd Street
while attending the funeral for
his uncle, Henry Clayton
Witmer in 1909

David Witmer was in college and headed to graduate school at Harvard to study
architecture, and would go on to an illustrious career as an architect designing,
among other houses and buildings, the Pentagon. The middle Witmer child,
William, would graduate from Harvard with a degree in engineering and go on to
assume control of the Witmer Brothers Company. Of the Witmer children, it was
William whose character most resembled that of his uncle, Henry Witmer (Fig. 3.8).

For Milt, the shock of his uncle's sudden death lingered as he made his way back to
the mountain. His persistence and determination when confronted with problems was
earning him respect, and with his eighteenth birthday approaching he was beginning
to plan for a move to Mount Wilson and a more permanent position as a muleskinner.
Completely at ease in his new life on the mountain, Milt sat down on a cluster of large
stones near his tent by the hotel and picked up a young red fox and laid the kit on his
lap, where it made itself comfortable. A rescue from the previous summer, the
orphaned kit was now in good health, Milt feeding it what he could from the hotel
kitchen and occasionally bringing home a live catch, hoping to teach the fox to
practice its hunting skills. Ever since his first visit to Mount Wilson Milt had learned
the importance of maintaining the balance of nature. Perhaps that is why he had come
to love his mountain home. On Mount Wilson, life seemed to be in perfect balance.
Even the founding of George Hale's new modern observatory was perfectly countered
by the naturalist tendencies of Hale and the staff at the facility (Fig. 3.9).

By the fall of 1909 Milt was working full time as a muleskinner. For his
eighteenth birthday his Aunt Alice gave him the beautiful black stallion he had
enjoyed riding on visits to the Witmer farm. The horse had been given the highly

Fig. 3.8 The family gathers at the Witmer ranch after the death of Henry Witmer in 1909. Milton Humason stands at the back resting his hand on a tree. William Witmer stands to his *left*. David Witmer stands to the *left* of William Humason, who is sitting on the ground at *right*. Joseph Witmer is seated at *left* near his cousin, Virginia Humason, who stands in a dress near the tent. Henry Witmer's widow, Alice Petterson Witmer, stands to the *right* of her sister, Laura P. Humason, at the picnic table

original name, Blackey, and Milt was often seen riding along the trail wearing chaps over his jeans and chewing tobacco as he peered out underneath his large brimmed cowboy hat. He moved Blackey to the stable near the base of the mountain so that he could care for him and took a room at the home of the stable manager, Chester Huston, at 215 Mira Monte Avenue. Milt's reliability had earned the respect of the slightly older stable manager, and the two had become friends.

As would be the case throughout his life, Milt made friends easily and had the trust of nearly everyone on the mountain. His knowledge of the surrounding hills was second to none, and his quiet confidence and cowboy charm and wit masked a penchant for mischief. He had become an expert horseman, fisherman and poker player, often taking his fellow cowboys for what meager pay they earned. A close knit group Milton and friends were prone to role-play, acting out robberies and holding each other up at gunpoint. An image from the period shows Milt dressed as a sheriff with a bushy mustache and a large star pinned to his shirt, leaning against the side of the hotel. Another image shows him squatting before the body of one of his cohorts who is lying face down on the hillside. The sheriff had apparently just shot the perpetrator. Although the guns were real no one was harmed in any of these

Fig. 3.9 Milton and the fox
he rescued on Mt. Wilson,
1909

Old West skits. In keeping with the traits commensurate with his station, Milt had developed a special gift for profanity that could be abrasive to the uninitiated. Long hours on the trail looking up the backside of a mule might have been the catalyst for these colorful imprecations (Figs. 3.9, 3.10 and 3.11).

Sitting with a group of his cowboy friends at a dance hosted by the hotel on a late summer evening in 1909, Milt was feeling very much a part of the picture on Mount Wilson. He sat taking in the scene as partygoers danced and sang accompanied by the small band assembled around the piano at the far end of the lobby. It was at this dance that Helen Dowd first walked into Milton Humason's life, making her way across the lobby floor whispering to her friend Nellie Campbell. Talking with his friends at one end of the room, Milt's attention was immediately drawn to Helen standing at the other end of the room. Charmed by her beauty, he couldn't resist asking Helen for a dance. As often happens in such cases it was love at first sight; the two connected instantly. After a few turns on the dance floor, they left the hotel and walked out to the perch overlooking Mount San Antonio where Milt had spent so many quiet moments over the years. There they sat for some time, taking their first awkward steps toward an uncertain future. It was the kind of night when the stars seemed to shine brighter than usual. But then Milt was color blind, so the stars always seemed brighter and sharper to him than most.

Fig. 3.10 Milt dressed as a sheriff at Strain's Camp on Indian Day, 1909

As they sat talking under the starry sky near Echo Rock Milt learned that Helen was actually the daughter of Merritt C. "Jerry" Dowd, the observatory's electrical engineer. A tall, rangy man with a steely gaze and a thick bushy mustache, Dowd had originally been hired to maintain and operate the truck that was being used by the observatory to haul cargo up the trail. He had a commanding presence and was often seen at the base of the mountain or on the trail coaxing the cranky machine up the hillside. Mule teams were used to help tow the truck along, and Milt and the other cowboys saw Jerry Dowd as one of the bosses at the facility.

Born Helen Joanna Dowd on July 23, 1891, in Pennsylvania, Helen Dowd had moved to Pasadena with her father and mother and her younger brother, Munson, in 1904. She was descended on her father's side from the Dowd family that had emigrated from England to colonial North America in 1639 and settled in Connecticut. Her mother, Katherine, who was three years older than her husband, was a second-generation American born in Pennsylvania to German immigrants. Jerry and Katherine Dowd had married young and moved from Pennsylvania to Massachusetts after Helen's birth. It was there that Munson was born in 1898. As a

Fig. 3.11 Blackey rears as Milt rides high in the saddle

result of its history, the Dowd family had taken part in the transformation of the country from British colony to vibrant sovereign nation. The family had a long tradition of education and was in good standing within the communities where they lived. Helen's grandfather, Julius Dowd, had written a book on grammar in the first half of the nineteenth century.

As fitted the times, Helen had grown up learning an appreciation of the outdoors. A girl of average height and slender frame, with an earthy fun-loving spirit and razor sharp blue eyes, she and her brother Munson had attended summer camp in Arroyo Seco near the family's farm home in the modern-day Simi Valley area. After her father began working part time as an electrician at Mount Wilson Solar Observatory years before, the family began making regular visits to the mountain. From the start Helen found the surroundings enchanting and visited her father at the observatory as often as her schoolwork and her mother would allow. The rustic surroundings and the buzz of activity both on the mountain and at the observatory and nearby hotel made Mount Wilson a favorite spot to visit.

After that first enchanted evening at the dance Milt and Helen began seeing more of each other. Photos that survive from those early years convey a spirit of friendship and good humor. On less formal occasions the young men could be seen in shirt and pants while the women wore ankle length skirts, button down shirts with neckties and colorful hats. On warm days in the summer Milt and Helen could be found wading in the West Fork of the San Gabriel River. In their quieter moments they would spend an afternoon together wandering their forest home, sharing dreams and making plans. In one photo, Helen stands near a large pine tree in a white short-sleeved blouse and a high waisted skirt and boots. She is holding in her left hand one of the ropes of a swing that hangs from the tree as she smiles broadly for the camera. In another photo Helen is dressed as an Indian woman with her hair in long braids for the annual Indian Day celebration on Mount Wilson (Figs. 3.12 and 3.13).

It was an exciting time to be on the mountain. The much-anticipated 60-inch telescope had come online, rendering its first images of the stars on December 20, 1908, just four years after the observatory's founding. Traffic up and down the mountain continued to increase as a night time observing staff, hired to operate the new telescope, and other visitors flocked to the mountain for a glimpse of the giant instrument. A museum built near the new dome and tower telescopes held images of the Sun and stars as well as information on local precipitation and the designs of the instruments themselves. Business was booming on Mount Wilson, and Milt and his fellow muleskinners were becoming one of its main attractions. They rode proudly in the Tournament of Roses parade, driving teams of mules along the streets of Pasadena, and stories of their antics and work on the trail were carried home by virtually every visitor to Mount Wilson. Poker was a favorite pastime of the cowboys, and they could often be found sitting on boxes and trunks around an old whiskey keg playing cards at the stable. The presence of the gun-toting trail hands gave the mountain the look and feel of the Old West.

Fig. 3.12 Tennis anyone? Helen Dowd and her best friend, Nellie Campbell prepare for a game

Fig. 3.13 Milt and his fellow cowboys sit on crates playing poker at the stable near the trailhead

With the large reflector now revealing never before seen images of the galaxy, Andrew Carnegie was planning a visit in the spring of 1910, and the observatory, plus the muleskinners who supported it, were hard at work preparing for the arrival of the institution's benefactor. The muleskinners rallied around Jerry Dowd and the truck, as longer sections of the 150-foot tower for the new solar telescope inched along the trail on their two-day trip to the summit from Pasadena.

As the rainy season arrived heavy work on the mountain had to be stopped to prevent any of the more precious cargo getting caught in a landslide while making its way to top of the mountain. On December 13, an overheated range caused a kitchen fire that leveled the dormitory known as the monastery. All available hands worked tirelessly to prevent the fire from spreading to the forest and other buildings. No lives were lost in the incident, and nothing of value was lost that couldn't be replaced. Work on a new dormitory to accommodate the visiting observatory staff began almost immediately in spite of the threat of landslides caused by winter rains. The new building, made of reinforced concrete, was complete and ready for Carnegie's March visit, and, as the rainy season was ending, work soon began to finish the 150-foot tower.

As he became increasingly aware that his daughter was falling in love Jerry Dowd informed Helen that that he in no way supported the match and that no daughter of his would be marrying someone with no future. The world was becoming increasingly industrialized. Ford automobiles were rolling off the assembly line at record pace, and the first car had made its way to the top of Mount

Wilson in 1907 (after which the driver said he couldn't be paid enough to attempt to do it again). It was only a matter of time before vehicles would be made that could be relied upon to handle the majority of the heavy lifting on the trail. The work of the muleskinners on Mount Wilson would soon come to an end, and Milt, who had no formal education or training past the eighth grade, would struggle to find work in the new world. If he wanted Helen's hand, Dowd said, he was going to have to find a job suitable to support her. For the first time since he had arrived at Mount Wilson years before, Milt's future seemed unclear. He was being made to choose between his life on the mountain and the hand of the girl he loved.

On a visit with the family in Crown Hill Milt discussed his problem with his parents in hopes of finding a solution. He lacked the training for a job at the bank that would suit his future father-in-law, and he had few options outside the family circle. His father mentioned the farm in Lordsburgh, and it seemed to be the most logical choice. He would feel more at home on the farm, and he could gain valuable experience in how to operate an orchard. If all went well he might someday have the means and the know-how to run an orchard of his own.

Since the death of Henry Witmer, the operation of the farm was being handled by Milt's cousin, Letha Lewis, a colorful character in her own right. The 40-year-old Letha was more than happy to have an able-bodied male and a member of the family to help run the place, and invited Milt out to begin working as soon as matters could be settled. For Milt the decision was clear; this was his best opportunity to get respectable work with a stable future. But the decision was far from easy. He had been living on or near the mountain for nearly five years and been a muleskinner for three of those. His entire future had been envisioned on the mountain, and now, in the blink of an eye, it seemed, that future had been altered. Putting this fact behind him, late in the spring of 1910, Milton Humason made his way from Mount Wilson toward a new life as a rancher in the nearby valley, at the foot of his former mountain home.

* * * * * * * * * * * * * * * * * * **

George Hale paced the floor of the hotel room he was sharing with his family in Paris. It was early 1911, and he had been touring Europe now for what seemed like an eternity. He was awaiting the arrival of the architect and longtime family friend, D.H. Burnham. The larger than life designer and urban planner was coming for a visit so the two could talk about the developments concerning the design and construction of the Hooker Telescope building and rotating dome. But the underlying reason was actually to take Hale's mind off of the events surrounding his beloved observatory, and Burnham was just the man to help him achieve this goal. Long a mentor to his old friend William Hale's only son, Burnham had always had a calming effect on Hale, who suffered more and more from the stress of running the observatory.

The heavy stress had started when the dormitory burned to the ground in December of 1909. At the time the observatory director had been busy planning the fourth annual meeting of the International Union for Cooperation in Solar Research, to be held in January of 1910. With the dormitory smoldering in ashes on the floor of the mountain, Hale and his assistant, Walter Adams, would have to quickly postpone the event until later in the year when accommodations could be made for the one hundred or so delegates. Hale wrote a hasty letter to the president of the Carnegie Institution of Washington asking for emergency funds to rebuild the monastery. The board at Carnegie immediately granted a special appropriation, and by the end of February the new building was standing.

On March 17, Andrew Carnegie made his way to Pasadena for a ceremony to celebrate the observatory and spend a night viewing the stars through the 60-inch telescope. Through the institution that bore his name, Carnegie had given tens of thousands of dollars for the building of the observatory, and now he had come to see for himself the fruits of his endowment. Among the dignitaries at the opening ceremony were John D. Hooker, benefactor of the 100-inch telescope, and the naturalist John Muir. Unfortunately for all who attended, the brilliant sunshine that blessed the opening ceremony turned to clouds and rain by afternoon, and Carnegie was forced to put his adventures in stargazing on hold for another visit.

By the end of Carnegie's visit Hale was near the point of collapse. He had put his heart, soul and mind into the observatory, pouring every last ounce of energy into the development of the land, the organization of the facility, its mission and leadership, the design, production and creation of its instruments and the funding for its operation. As a result the Mount Wilson Observatory was on its way to becoming the greatest facility of its kind in the world, with two new powerful solar telescopes and the world's largest fully functioning reflecting telescope. These instruments were beginning to reveal images of the Sun and stars never before seen, and new discoveries such as the Sun's magnetism, which Hale had discovered in 1908 around the time the 60-inch telescope was seeing first light, seemed imminent. It had been the greatest adventure of Hale's life, and the most taxing. His heart and soul were still up to the task, but his mind was giving out. Friends, including Huggins and Burnham, began receiving disturbing news of sleepless nights and constant and severe headaches. Lately an imaginary little man had begun visiting the 43-year-old director and was becoming a permanent fixture in his daily life.

The central problem driving Hale's anxiety was the creation of the glass blank for the 100-inch reflector. The first attempt had been deemed useless on its arrival to the shop in Pasadena, and the second disk, which had been buried in manure for months during annealing, had developed stress points and fractured. Now a third attempt to cast the enormous disk was in the works. Although he was supposed to be resting and not paying any attention to developments at the observatory Hale couldn't resist visiting Mr. Delloye at the French Plate Glass Works while in Paris. Despite reassurances to the contrary a successful casting of a disk of the size needed to create the 100-inch mirror seemed in doubt. To make matters worse, Ritchey, who had been in private conversations with Hooker, had designed a composite mirror and was pushing the merits of his design in lieu of the single glass disk. To Hale's eternal dismay, Hooker had begun to buy into Ritchey's scheme and was growing ever more gloomy about a successful outcome to Hale's original plan.

Ritchey's betrayal had incensed Hale, and in reaction to the insubordination the director had reduced his chief optician's role at the observatory. But the damage had been done. When the delegation finally arrived in late August of 1910, winding its way up the mountain trail in the summer heat in a local media frenzy, Hale was barely able to attend the conference he had been so looking forward to. Embarrassed and dejected Hale soon left for Europe with his family to try to recover his health. The group first sailed to London, where Hale had enjoyed time reading and relaxing in the halls of the Athenaeum. But then a letter from Hooker arrived, saying that all work on the 100-inch project had ceased until the viability of the casting of the mirror blank could be assured. This letter prompted Hale to visit Paris and speak to Delloye at the glass works in Saint Gobain. While Hale was in Europe Hooker had visited Walter Adams, who was acting as director of the observatory in Hale's absence; Hooker was trying to bully the assistant director into signing a release excusing him from any further expenditure on the telescope's behalf. Adams signed nothing and wrote to Hale asking him to advise him on how to proceed. The news had sent Hale to the edge again, prompting his wife, Evelina, to write to Adams angrily, "I wish that glass was in the bottom of the ocean."

While in Paris reeling at the word coming in from Pasadena, Hale finally had a bit of good news. Adams had been testing the first disk at the lab and found it to be sound. The disk had been poured in three layers, and Adam's finally concluded that the combined layers acted as one complete unit. This, he decided, rendered the disk suitable for grinding into a paraboloid mirror. Heartened by any good news regarding the ill-fated disk, Hale immediately wired Adams to begin the work of grinding the disk.

This event preceded the arrival of Burnham in Paris by a few weeks and set the table for a discussion of the telescope building's design. The future home of the great telescope had been designed by the famed architect, who had achieved world renown as the chief designer of the Chicago World's Fair in 1893. Burnham, a good friend of Hale's father, had designed several Hale family buildings, including the sprawling castle-like mansion in the Hyde Park section of Chicago. When George was ready for college, Burnham had recommended M.I.T. to his father as an ideal place for an advanced degree. Paying it forward, Hale had put his energy behind the transformation of Throop Technical College into the California Institute of Technology. Caltech was being created to compete with Hale's alma mater for primacy in educating elite students in science and technology in the West. Burnham had served as a guide, aid and mentor to George throughout his life, and his support and creativity helped calm emotions concerning the 100-inch project.

Burnham's arrival renewed Hale's health and humor, and the two spent several days visiting local architectural and civic sites. With the bubble-ridden disk being shaped in the observatory lab in Pasadena progress seemed finally to be inching forward. Word of the disk's viability had eased Hooker's anxiety for the moment, and Hale knew that if he could secure funds to build the dome and mounting the telescope's future would at last be realized. That dream was granted when Andrew Carnegie announced that he was giving another $10 million to the Carnegie Institution of Washington, with funds to be allocated to the completion of the

Hooker Telescope and dome on Mount Wilson. Unfortunately, news of the endowment had not prevented Hale's unsteady health to decline, which prompted him to spend even more time away from Pasadena. He sent word to Adams that he and the family were setting sail for Egypt, where he would have little or no contact as he tried to regain his health once again. As he and his family set sail for Cairo, the future of Hale's beloved observatory seemed secure, but his ability to contribute to its daily life was in grave doubt.

* * * * * * * * * * * * * * * * * **

On October 15, 1910, William and Laura Humason sailed into the harbor in San Francisco aboard the U.S. *Mongolia*. They had just returned from a long tour of China and Hong Kong in celebration of their twentieth wedding anniversary. On their way home they stopped in Honolulu for several weeks before boarding the cruise liner (a converted cargo ship) for the West Coast of the United States. The much anticipated trip had been postponed so that the family could relocate and refocus after the death of Henry Witmer. While his parents were away, 18-year-old Lewis was left in charge of the household. On their return the family got together to hear the stories of their trip abroad.

Milt sat quietly listening to his parents as they told the tale of their tour abroad. He had always enjoyed a good story, but on this occasion his mind was elsewhere. For the past few months he had been learning the ropes at the ranch. Although they were cousins by marriage, Letha Lewis was only a year younger than Milt's parents. She was the oldest child of Samuel and Mary Agnes Lewis. Despite their age difference, Milt and Letha shared a love for theater and the outdoor life. Now in her early forties, Letha's charisma and character made her the life of the party during frequent outings with her friends. An early advocate for equal rights Letha had married Sam Storrow, a mining engineer from an established Massachusetts family, in her thirties. The exceedingly progressive society woman probably accepted Storrow's marriage proposal to secure her standing within the circles she was traveling. She had a strong will and a rebellious spirit and could often be found driving her automobile wearing riding a riding hat and goggles or trapshooting in a dress with a group from one of the many clubs she supported. She loved plays and poetry and loved to entertain with her favorite dramatic passages.

As jobs went, life on the farm was pretty good, and Milt took to the work with his usual sense of pride, easily learning the tricks of the harvest and trade life of a rancher. The farm had always been a home away from home, and the ranching life felt very close to his work on the Mount Wilson. The stable on the ranch also gave him a place to look after his horse. But his eye was always on marrying Helen, the girl of his dreams. During the summer months the two would often go for afternoon rides through the countryside on horseback. For a laugh, Milt would sometimes saddle the dairy cow and ride it around the corral.

Halley's Comet was revisiting Earth on its 75 year journey around the Solar System. While the comet, with its brilliant tail, was still visible in the night sky,

Milt and Helen would sit out and marvel at the site of it streaking across the heavens.

Whenever they had the chance they would join their friends on Mount Wilson for one of the frequent events that were sponsored on the mountain, including the occasional public nighttime viewing of the stars through one of the telescopes at the observatory. Milt marveled at the incredible magnifying power of the instruments, and he and Helen often opined the loss of the life they had enjoyed on the mountain they loved. They both agreed the change was worth it and planned to marry as soon as Milt's position on the ranch improved enough to convince Jerry Dowd that Milt was worthy of his daughter's hand. As they closed the first decade of the twentieth century Milt and Helen were dreaming of making a life together.

Chapter 4
Life in the Valley

Abstract After proposing marriage, Milt and Helen marry and move to a family farm to take up a life of ranching. Milt's mother and aunt go over seas with Virginia and Joseph and get caught in the early stages of World War I. Milt and Helen give birth to their son, Billy, in 1913 and settle into a ranch of their own in 1917 as the U.S. gets deeper into the war effort. About the same time, the 100-inch telescope is completed opening up a position on the mountain for a janitor and part time night time assistant, that Milt decides to take.

After a year on the "Lewis" ranch Milt had become familiar enough with the details of the operation to assume the position of foreman. His job included seeding and nourishing fresh groves, tending the existing orchards and scheduling and planning the harvest of each crop. In addition to oranges, the farm also yielded apples, peaches, figs and almonds, each with its own ripening cycle and treatment needs. The ranch was also the home of cattle and horses, which Milt gladly tended to as part of his daily routine. The added value of life on the farm was that it was a family affair. When the county fair or some other event came around members of the Humason, Witmer and Lewis families were sure to descend on the farm for a weekend of picking and packing the fruit and nuts to be shipped to the event. A ride or a hike in the back country was always a welcome diversion, and there was plenty of fruit pie and preserves to go around. The commitment to help out extended to the rest of the Witmer business interests as well. By now Henry Witmer's sister, Anna Victoria, had assumed control of the Witmer Brother's Company in Los Angeles, and Henry's widow, Alice, was now the president of the L.A. Improvement Company, which dealt in the family's land holdings. William Humason was acting as secretary at Witmer Brothers (Figs. 4.1 and 4.2).

For Milt, having the family around was a comfort as he tried to gain his footing in his new life. After years on Mount Wilson, adjusting to life in the valley was a big change. The starry-eyed faces of visitors and researchers making their way up to the observatory for a tour of the grounds, buildings and telescopes had been replaced by the humdrum daily assortment of farm hands, distributors and retailers. On the mountain, every sunrise brought the promise of a new discovery. In the valley, only the frequent visits from his family and the promise of a life with his sweetheart, Helen, motivated him.

© Springer Science+Business Media New York 2016
R.L. Voller, *The Muleskinner and the Stars*,
Springer Biographies, DOI 10.1007/978-1-4939-2880-4_4

Fig. 4.1 Milton Humason
with his father, William Grant
Humason, 1910

And so it was that a little more than a year after he left his job as a muleskinner
on Mount Wilson to secure a job that would appease Jerry Dowd, Milt approached
him to ask him for his daughter's hand in marriage. This time Dowd was more than
willing to give his consent. On a pleasant autumn night while sitting and gazing at
the stars from "the perch" on the summit of the mountain, Milt proposed to Helen,
and she accepted.

They were married, amid close family and friends, in a small ceremony at the
Dowd's home on Elevado Drive in Pasadena, October 14, 1911. Milt looked
handsome in a black suit coat and pants with white shirt and tie, while Helen wore a
simple but beautiful white dress and broad-brimmed hat. Helen's friend, Nellie
Campbell, and Milt's cousin, Ralph Witmer, acted as maid of honor and best man
for the couple. Afterward, their entourage made its way to the ranch for a leisurely
afternoon of fun and celebration. Photos from the day show Helen walking with
Stella, the collie mix who lived on the farm, and Milt atop the mighty Blackey. One
of the mares had recently given birth, and Milt and Helen had their photos taken
while standing with her foal in the corral. These were scenes right out of the Old
West!

With his marriage to Helen Dowd, Milton Humason, at age twenty, was once
again showing the courage of his convictions, having turned another dream into

Fig. 4.2 Milton LaSalle
Humason dressed as a
cowboy

reality. What's more, he had made a trusted ally in Helen's father, Jerry Dowd, who would figure prominently in future events. For now, though, the two settled into a simple life on the ranch in La Verne (Figs. 4.3, 4.4, 4.5 and 4.6).

Milt took to farming the same way he had to the cowboy life on the mountain, with a special pride of ownership and sense of responsibility that had made him one of the most respected of the muleskinners on Mount Wilson. As foreman of the ranch now being operated by Sam Storrow and Letha Lewis, Milt had to manage all aspects of the output and maintenance of the orchards. The ranch's main yield was from fruit-cropping, but it was also home to horses, chickens and a dairy cow. Every day, as he worked in the fields and orchards around the ranch, Milt could look up to the top of Mount Wilson for a glimpse at his former home. The two solar towers could be seen above the trees, and the large dome of the 60-inch telescope stood on a nearby hilltop, its great white crown shining in the midday Sun. Although he was becoming very much at home in the valley, Milt's heart was still on top of Mount Wilson.

On November 5, a few weeks after Milt and Helen were married, Calbraith Perry "Cal" Rodgers landed his Wright Model EX airplane in Pasadena. Rodgers had just completed the first transcontinental flight, having taken off from Sheepshead's Bay, New York, on September 17, flying south to Texas from Chicago before turning west. Rodgers had been attempting to reach the West Coast in thirty days.

Fig. 4.3 Helen stands near a tree swing on Mt. Wilson in 1909

Considering the condition of both Rodgers and his airplane when they landed it is doubtful the train or cruise liner industries took much notice at the time. The harrowing trip took Rodgers 49 days to complete, and he suffered sixteen crashes nearly killing himself several times along the way. His actual flight time was less than four days, and when he landed only a few parts remained from his original flyer.

Despite these many setbacks Rodgers feat showed the incredible innovations being advanced by technology. Telescopes at the observatory were trained on the southern mountains for a glimpse of Rodgers plane on the day of his arrival. The event turned the town on its head, and a crowd of 20,000 turned out to greet Rodgers when he landed at Tournament Park, the site of the annual Rose Bowl

Fig. 4.4 Milt standing near a burnt tree after a fire wearing his Mountie hat

Fig. 4.5 "The perch" on Mt. Wilson where Milton Humason proposed marriage to Helen Dowd in 1911

Fig. 4.6 Milton and Helen Humason on their wedding day, October 14, 1911

parade. The crowd rushed the plane, draping the aviator in an American flag before driving him through the city to a hero's welcome.

For the new couple, being immersed in the family fold brought with it certain luxuries of this increasingly modern age. In a photo from the time, Milt and Helen laughed it up for the camera as Helen stood holding the hand crank on the front of Letha's Ford Model T. Slender and thin-waisted, she wore an ankle-length white dress with sleeves to her elbows and an exotic broad-brimmed feather hat. These hats would begin to go out of style after about 1912, but Helen looked quite comfortable in hers. Standing beside his new bride, Milt cut a handsome figure peering out under the brim of a white fedora. Dressed for summer, he wore a white short-sleeved shirt and tie with white pinstriped pants. Headed out for a drive or to some other function, the two bore the look of comfort and leisure (Fig. 4.7).

* * * * * * * * * * * * * * * * * **

Visits to the observatory were fairly common for Milt and Helen. For convenience a motor stage now ran from Pasadena to the summit of Mount Wilson, making the 9-mile journey in about two and a half hours. The old familiar faces were still visible, but some new ones were being added to fill in the staff of stellar and solar

Fig. 4.7 Milt and Helen clown around near Letha Lewis's car at fairgrounds in Pasadena

observers. The latest addition was Adrian van Maanen, a recent graduate of the University of Utrecht. The completion of the 150-foot solar tower lent a new attraction to visitors of the observatory. The tall solar telescopes were constantly tended during the day by the colorful Ferdinand Ellerman, one of the original five members of the staff on Mount Wilson. Sporting a bushy mustache and van Dyke beard with round tortoise shell eyeglasses, Ellerman bore the look of an English gentleman, but he was always on the lookout for an opportunity to entertain visitors and staff alike. A favorite diversion of his was to hop into the elevator of the 150-foot solar tower (known as the "bucket") on a sunny day and ride it to the top of the tower while a crowd had gathered beneath him.

To the uninitiated the ride to the top of the tower was a scary proposition, but Ellerman showed no fear. With the ease of a man standing on a street corner, he would lean back in the bucket while filling his pipe with tobacco, and ride half way to the top of the tower before stopping. Pulling out his pipe and a magnifying glass, he proceeded to light the pipe using the glass to concentrate the heat from the Sun on the tobacco to light it. As he puffed the pipe, Ellerman would then start the elevator and continue to the top of the telescope with the crowd cheering below.

Things were going well on and off the mountain for the observatory. George Hale was finally back working in his capacity as director after a 16-month leave of absence due to illness. In May of 1911 John Hooker had died, leaving his debt for the 100-inch telescope unpaid. Fortunately for Hale and the observatory, the influx of cash from Carnegie made funds available to finish the work. Although Hale had forced his retirement from the observatory George Ritchey remained in charge of the grinding and polishing of the mirror blank at the lab in Pasadena. At

8000 square inches polishing the mirror true to within millionths of an inch was a monumental task. News of the numerous failures to cast a satisfactory disk by the glassworks in Saint Gobain had understandably given Hooker cold feet, but Hale knew that if the mirror being polished in the shop were to become viable the telescope could be built. If polishing the glass disk did appear successful, the plans for the building and dome had been completed by Burnham and were ready for construction.

The new office building was being built right beside the lab in Pasadena, with two upper floors for offices and a library on the main floor. The Spanish-style building featured a concrete reinforced basement where the plates now being brought down from the mountain almost daily could be stored.

As growth at the observatory escalated and newer technologies were needed, Jerry Dowd found himself more a part of the development at Mount Wilson. The newer equipment required a bigger DC generator, and buildings needed to be outfitted with the latest in electrical wiring and switches.

* * * * * * * * * * * * * * * * * **

In the months and years following her husband's death, Milt's aunt, Alice P. Witmer, had tried to put the loss behind her. After grieving for a year or so she decided to launch herself into work on the family business, taking over as president of the Witmer Brother's real estate venture, the L.A. Improvement Company. But neither time nor hard work were enough to get her over the loss, and she asked her sister, Laura, to go with her on a long trip to Europe. Together they could see some sites, and the sisters believed the change of scenery might do her some good. They planned to leave as soon as William returned from a business trip to the British territories in Hong Kong in June of 1913. The oceanliner, *Titanic,* had sunk in the North Atlantic in April of 1912, and the general mood of the country toward open sea voyages had dampened in the aftermath of the great ship's sinking.

Nevertheless they decided to take their chances, and in the summer of 1913 they boarded a ship for Paris with Ginny and Joseph in tow. After touring France and Switzerland they decided to go to Rome and tour Italy. Needing some free time, Alice and Laura decided to enroll Ginny and Joseph in boarding school in Montreux. With the children safe the sisters made their way south for a tour of the Italian coastline. It was there that Laura Humason received some surprising and wonderful news. In her absence she had become a grandmother. On October 30, 1913, Milt and Helen brought their only child, William Dowd Humason, into the world.

With mother and baby doing well Laura decided that she would stay with Alice for a while longer before returning to Los Angeles the following year. The sisters went on with their tour until, on June 28, 1914, the Archduke of Austria was assassinated in Sarajevo. Immediately plans changed, and Laura and Alice headed back to Switzerland to get Joseph and Ginny and return home. Their plan was halted, though, by the invasion of Belgium and Luxembourg by Germany in July. This caused Britain to declare war on Germany, and before long the German positions had been pushed back and a stalemate devolved into trench warfare along what came to be known as the Western Front. Fearing for the children but heartened by Switzerland's neutrality in the conflict, Alice insisted Laura return to her family in Los Angeles, telling her she would stay behind and retrieve the children as soon as she was able. With uneasy feelings Laura did as her sister advised, boarding the S.S. *Cameronia* from Glasgow on August 15, 1914, bound for New York. After some weeks of traveling, first by boat and then by cross country rail, Laura was finally home in Los Angeles cradling her grandson in her arms.

In the meantime, Alice waited for word that it was safe to return to Montreux to collect Ginny and Joseph, who were now 16, which she did in the spring of 1915. With war erupting across Europe, Alice was in a hurry to get the children back safely to American soil. On April 29, emergency passports were issued for both Ginny and Joseph in London, and Alice made reservations for their return trip home aboard the S.S. *Lusitania* for June. On May 7, the *Lusitania* was struck by a German torpedo and sank on its way to Liverpool from the United States, killing nearly 1200 people. The shock of the news was little deterrent to Alice. She made another reservation for September of that year and sent the Joseph and Virginia back to boarding school to keep their minds off the hostilities. Finally, in September the trio boarded the S.S. *St. Louis* at Liverpool, and, after eight nail-biting days on the open seas, they arrived safely in New York harbor on September 12, 1915.

By late September the family were together again, exchanging stories on the events both at home and abroad. Much had changed since Laura, Alice, Joseph and Ginny had embarked on their journey. In 1914, William Sullivan Witmer, son of Joseph and Josephine Witmer, had arrived in Los Angeles, and he had since taken over as secretary of the Witmer Brother's Company. Milt's father was more than willing to step aside and let the Harvard educated heir to the Witmer legacy assume his new role in the family business. The 24-year-old moved in with his aunt Anna Victoria at the Lewis family home on 1425 W. 3rd Street in Crown Hill. William's older brother, David Julius, was studying architecture at Harvard and planned to join his brother in California when he completed his studies. The presence of the avuncular William Witmer and his older brother brought a youthful and vibrant energy to the families living in Crown Hill (Fig. 4.8).

However, for the moment it was baby William who was stealing the show. The year "Billy" was born Milt and Helen moved to a new home on 534 Palmetto Drive, while Milt continued his work as "gardener" for Sam Storrow. Infant photos show a smiling blond headed boy standing on a chair wearing a white suit and knit hat, sitting on his mother's lap on the porch of the family home, and playing on the foot rail of his Aunt Letha's car. Milt's penchant for dress up and practical jokes meant there was always fun to be had at the Humason home. Another photo from the period showed 3-year-old Billy standing on some rocks near a river holding a small fishing pole with a creel draped around him. The wicker basket, which fishermen used to hold their catch, is shown dangling at the youngster's feet and was nearly as big as the boy wearing it! Whether he was Billy the boxer or Billy the fisherman or Billy the nascent gardener, there was always a laugh to be had as long as his father was around. On a family picnic Milt has his picture taken while cutting a loaf of

French bread with an axe over a log, a primitive but effective means of slicing bread! (Figs. 4.9, 4.10 and 4.11).

There was much fun to be had outside the family home as well. The motion picture film industry was introducing its newest star, Charlie Chaplin, to the public. Chaplin introduced his iconic character, The Tramp, to wide acclaim in his second released silent film in 1914, and two years later he was an international superstar. The actor turned director and producer lived and worked in Los Angeles and was at least once the house guest of Letha Lewis and her husband, Sam Storrow.

While the birth of Milt and Helen's son breathed new life into the family, the return of Josephine Witmer and her children to Los Angeles in 1915 and 1916 was a heartwarming victory for all. By now everyone within the family circle were acquainted with the story of the hardships the Witmer women and their children had endured and the occasion of their being reunited was cause for celebration. In 1915 William and Laura Humason moved to a house at 986 Fedora to make room for the arrival of Josephine and her daughter Mary, who moved into Alice's home on 1422 West 3rd Street. Jo, who was five years younger than Alice, assumed the head of the household. David Witmer, the aspiring architect, soon after arrived in Los Angeles and moved in nearby with his younger brother William at 1207 West 3rd Street.

Fig. 4.9 Billy sits on a steel girder during the construction of the 100" dome in 1915

Near the end of 1916 Milt and Helen were getting ready for another move, this time to a ranch of their own. In the five years he had spent on the Lewis ranch working for Sam Storrow, Milt had been saving his money for the purchase of land when good soil could be found at a reasonable price. He found a ranch at 225 Hillcrest Avenue, a mile or two from the Storrow ranch, and after perusing the land and negotiating a price, bought the property. The centerpiece of the property was a beautiful late nineteenth century two-story ranch house. A large palm tree stood outside the broad front porch, and the new home had more than enough space to accommodate Milt and Helen's parents on their frequent visits. Milt was gaining confidence as a rancher and had good connections within the ranching community. He took to the work on the new ranch with a sense of pride and purpose that only self-ownership can produce.

* * * * * * * * * * * * * * * * * **

Living life in the valley was increasingly taking Milt's mind off of Mount Wilson. As his responsibilities, at home and at work, grew his only real point of contact with the observatory was his father-in-law, Jerry Dowd.

Fig. 4.10 Boxing Billy near the Humason's home on the Lewis ranch in 1915

In the years since the young Humason couple moved away Jerry and Katherine Dowd had moved into a little ranch on the mountain. His skill as an engineer had made Jerry an integral part of the developments in the growing observatory. On the many visits to their grandson at the Humason home, Jerry would fill Milt in on events at the summit. Much press had surrounded the burning of the Mount Wilson Hotel in 1913. In the aftermath of the fire the hotel's owner, the Mount Wilson Toll Road Company, built a much bigger and more modern facility, complete with outdoor pool, which opened to the public the following year.

Several buildings and small cottages had been built to accommodate the growing science and support staff. The road to the summit was much wider now, but still, as Milt learned, the road to the top with some of the larger pieces of the new telescope had been an adventure. During an ascent the road gave way under the rear wheel of the 3-ton truck carrying one of the tube sections and had to be lashed to the mountain with ropes and chains until a mule team could be brought into help tug it to safety. As usual work on the mountain brought with it harrowing stories of calls too close for comfort. But the center of attention was the anticipated completion of the 100-inch Hooker Telescope. The foundation and pier had been finished in the summer of 1914 before the winter rains set in, and the building and dome were constructed starting in 1915.

Milt and Helen had visited the grounds with Billy while the dome was under construction, and Milt took a photo of the boy, seated on the skeletal frame of the

Fig. 4.11 Generations: Milt stands beside Billy who is sitting on a mule near the halfway house on the Mt. Wilson trail. Virginia Humason is at right while Lewis Humason stands to the left of his cousin, Joseph Witmer

dome, to commemorate the occasion. As the dome and mounting were nearing completion Dowd was installing new push-button controllers, pumps and other systems in the Hooker dome as well as the rest of the facility. A new spectrographic camera, one-tenth the size of the Hooker mirror, had been installed in a small dome nearby (Figs. 4.12 and 4.13).

Complications in designing, building, transporting and erecting the Hooker Telescope had more than once driven George Hale to the breaking point. The U.S. involvement in World War I would soon resurrect him. As the United States grew closer to entering the war in Europe, the respected observatory director was called to Washington, D.C., to organize the National Research Council with special interest in aiding Britain and her allies. By the second decade of the twentieth century the United States had become an industrial powerhouse. Ford Motor Company had rolled its one-millionth Model-T off the assembly line in 1915, and another engineering marvel, the Panama Canal, was completed in August of 1914, opening shipping lanes between North and South America. For the first time ships carrying cargo and passengers to either coast of the Americas could do so without having to sail all the way around the southern tip of South America. The new canal, widely seen as one of the manmade wonders of the world, cut through some 50 miles of land in Central America at the Isthmus of Panama. The United States bought the canal, in

Fig. 4.12 Who wants bread? Milt clowns around during a picnic in the forest on the family ranch

Fig. 4.13 Billy's first catch!
Fishing the San Gabriel River,
1916

1904, ten years after the French government had dropped the project due to the high cost and an alarmingly high death rate within its labor force. Once finished the complex system of locks raised ships up to the desired height of 85 feet above sea level. The completion of the canal was one of the greatest engineering feats in history and was seen as another prime example of American ingenuity, building prowess and resourcefulness. These were contributing factors to Germany doing what it could early in the war to keep America from entering into the conflict.

After the sinking of the *Lusitania* by a German submarine in May of 1915 public sentiment in the United States began to turn toward getting into the war. Although Republican figures such as Theodore Roosevelt supported U.S. involvement, President Woodrow Wilson called for restraint, demanding Germany stop unrestricted attacks on mercantile ships in international waters. At the time Germany heeded the demand in an effort to cool tempers in the United States. Still news of the horrors of the battlefront was unsettling to most Americans. Trench warfare and other outdated military tactics clashed with the major innovations in weaponry of the age to devastating effect on both sides. These factors and the continued U.S. abstinence from the war led Wilson to reelection in 1916. The Princeton-educated president, who had a Ph.D. from Johns Hopkins University, was one of the most progressive voices in American politics since Abraham Lincoln.

Even so, Milt, who had grown up as a rock-ribbed Republican, didn't fancy a Democrat in the Oval Office during a time of war. In January of 1917, Germany resumed unrestricted submarine warfare, sinking merchant ships crossing the Atlantic. The German leaders knew the move would bring the United States into the war but believed it would be many months before the arrival of American forces. This was a gross miscalculation. On April 6 America declared war on Germany, and the Selective Service Act, which went into effect in May, would grow the U.S. military by almost 3 million by the following summer.

As their loved ones signed up for the draft and went off to war, citizens of the United States braced for the worst. Milt and Lewis went down to the draft board together with their cousins, David and William Witmer, on June 5, 1917. On his draft card the 25-year-old husband and father listed his employer as "self" and his occupation as "rancher," along with the address of the family farm on Hillcrest Avenue in Monrovia. Lewis, a mechanic who had studied at the University of California at Berkeley, was soon drafted into the army and served in an armory battalion during the war. David Witmer would be drafted as a captain in the army the same year after marrying his sweetheart, Helen Williams, whom he had met in Massachusetts. For the duration of the war, David's new bride lived with his mother and aunt in the family home at 1422 West 3rd Street. William Witmer, who was never drafted, spent the war managing the Witmer estate while helping to run the Witmer Brothers Company at the Wright Callender Building on 405 South Hill Street in Los Angeles.

* * * * * * * * * * * * * * * * **

As the country prepared itself for war the observatory was preparing for the arrival of the 100-inch mirror, which was in the final stages of shaping and polishing at the lab in Pasadena. The momentous occasion was set for the Fourth of July weekend and would be visited by throngs of supporters, onlookers and reporters. As the telescope's big day approached Milt and Helen got a surprise during a visit with Helen's father. Jerry Dowd and the staff at the observatory were entering the final phase of assembly and development of the 100-inch telescope, believed to be coming online in the very near future. New facilities were being built to house lab equipment, grounds-keeping supplies, and more. In the wake of the new growth, staff were going to be needed to help work and maintain the telescopes and facilities. Dowd had learned that the position of janitor was opening up and said that the job offered Milt an opportunity to help out as a part-time night assistant on the big reflecting telescopes. The thinking at the time was that the new reflectors and their companion spectrographs might eventually lead to longer and longer exposure times as spectra of fainter objects were pursued. The night assistants would be trained on the operation of the telescopes so they could help out the staff astronomers during longer observing runs. The operation included learning how to change out the focus cage of the telescopes, maintaining the equipment and dome, and positioning the telescope.

Fig. 4.14 The Humason ranch in Monrovia, 1918

Fig. 4.15 Horse and plow. Milt takes a break from his work for a photo

Ordinarily the offer to leave his life as a self-employed rancher to take a job as a janitor would be a short conversation. But the situation on Mount Wilson had changed a little. First of all, the chance to live and work on Mount Wilson again was enticing. Both Milt and Helen loved the mountain, and in his short life Billy had learned to love it, too. For Milt, the entire enterprise hinged on his ability to work with the large reflectors. Linked as it was to the Carnegie Trust, a job at the observatory did offer a stable living for the young family, and what better environment for Billy to grow up in than a facility full of some of the world's top scientists. Although the idea was intriguing Milt and Helen decided to hold onto their ranch for a year until they were sure the move was in their best interest. Milton Humason was concerned about leaving behind a secure and respectable life in the valley to again face an uncertain future on the summit of Mount Wilson (Figs. 4.14 and 4.15).

Part II
Ordinary to Infinity (1917–1938)

The clearest way into the Universe is through a forest wilderness.

John Muir

Chapter 5
Winning Friends and Favor

Abstract Having been hired by the observatory as a janitor and part time night assistant, Milt quickly makes friends and soon finds he has a gift for stellar photography. With the help of friends like Seth Nicholson and Hugo Benioff, he begins to learn the craft of stellar spectroscopy and the physics that supports it. As his skill at the telescope grows Milt publishes his first reports on cosmic events.

On his return to Mount Wilson, Milt encountered a world far different from the primitive wilderness outpost he had confronted fifteen years prior. Buildings and domes and towers and cottages seemed to be everywhere. The hotel that replaced the one that had burned years before was lavish in comparison and much larger than its predecessor. Life at the research facility on the hilltop above Pasadena was increasingly vibrant and intriguing. The mountain was evolving, and Milt, who had come back to Mount Wilson as a young and successful rancher, husband and father, was evolving, too!

Meanwhile, in Europe and Asia, the world was dissolving into chaos, and the effects of the ongoing conflict were being felt at home. In his opening address to the Contribution of the Mount Wilson Solar Observatory for the Carnegie Institution of Washington Yearbook in 1917, George Hale indicated the seriousness of the effects of the war: "The far-reaching influences of the war, so disastrous in their effect on the progress of science in Europe, are clearly reflected in the observatory's history for the past year."

The director was referring chiefly to the absence of Jacobus Kapteyn, a Dutch astronomer well known among astronomers for his work in studying the stars of the Milky Way. In 1906 he had begun the first collective study of the sky in history, coordinating a detailed analysis of the distribution of stars in the galaxy between dozens of observatories worldwide. A visiting Research Associate of the Institution since 1907, Kapteyn had been unable to leave his home at the University of Utrecht in Groningen for the past year as war spread across the European continent.

Another casualty of the observatory from the war was Dr. Arnold Kohlschötter, who had been captured by the British while trying to return home to serve in the German army in 1915 and would serve out the remainder of the war as a British POW.

* * * * * * * * * * * * * * * * **

With Hale away most of the year helping to manage the war effort, assistant director Walter Adams was left in charge. Few events in the history of the Mount Wilson Observatory rivaled what occurred on July 1, 1917, the day the 100-inch mirror finally reached the summit. There was so much hype surrounding its arrival that a large crowd formed near the entrance to the toll road that morning. Photographers lined the route, hauling their cameras up the trail for a prime shot of the massive mirror making its way to the summit. Everywhere tiny American flags waved in the breeze, attached to the windshields of motorcars, sprouting from men's and women's hats and stuck to the side of the station house. The arrival of the great mirror on the eve of the celebration of the country's independence made it an occasion to be remembered. Women dressed in skirts and bonnets and men in suit coats and boots waited at the bottom of the hill to accompany the mirror on her journey to the top of the mountain. Muleskinners were busily preparing their best mules to aid in the mirrors ascent should they be needed.

The 100-inch mirror stood on edge in a wooden crate that was bolted to the bed of the Mack AC model truck, one of two trucks the observatory had purchased for hauling heavy materials up the mountain. The trucks were being housed in a garage on the campus of the Pasadena offices. Sturdy wooden bracing, installed on top of the flatbed of the truck, kept the mirror secure, while springs placed between the crate and the bracing reduced vibration and smoothed the ride during the trip. At rest near the toll house station at the foot of the dirt roadway, shielded from the hot Sun by a white canvas tarp, the mirror stood 14 feet above the ground and the entire load including the bracing weighed almost 8 tons. The new truck had a dash-mounted radiator under its distinctive snub nose hood cover and a chain-driven rear-axle that allowed it to power its way up steep terrain. Despite the truck's size and improved hauling ability the steep slopes of the mountain and the fragile nature of its cargo required a full day's work with the help of 200 men to tow the heavy payload to the 100-inch dome on the observatory grounds.

Jerry Dowd, who had been busy most of the previous year designing, building and wiring the complex switchboard and control apparatus for the telescope, was on hand to drive the truck up the hillside. It was a day filled with joy and frustration as the truck made its way slowly up the mountain. Everyone associated with the observatory was on hand to see for themselves the new marvel of science make its way toward history. Finally, ten hours after it had left the base of the mountain, the truck arrived with the mirror intact. At last the great telescope, comprising more than 700 tons, was nearing completion.

* * * * * * * * * * * * * * * * * **

In the fall of 1917 Milt and Helen took up residence in a small cottage on Mount Wilson. Set into the hillside beneath a grove of beech and linden trees the wood-sided bungalow featured several small rooms and windows with white outer trim under a gabled roof. The mountain offered few of the conveniences and amenities of the life they had led in the valley. Apart from the daily delivery of the mail by stage, running water and electricity were about the best that could be hoped for.

The harshness and unpredictability of the climate on Mount Wilson was part of the reason George Hale had designed it to accommodate part time and not full time staff. Summer brought the threat of forest fire, and, sometimes, torrential rains would wash out the road to the city. A winter snowstorm could dump enough snow on the peak to strand residents at the hotel, observatory and nearby houses, cutting off access to food and other supplies. Other dangers lurked as well. Although the Humasons shared a love for wildlife, the fear of bears, mountain lions and snakes was heightened by the presence of young Billy. Still, to Milt and Helen, the risk seemed worth the reward. For them Mount Wilson was a magical place, the place where they had met and fallen in love. Having the chance to raise their son against that backdrop while taking part in the historic development of the observatory was more than enough compensation for the harsh realities of mountain living. For four year old Billy, the adventures he would share with his father in the years to come would be retold throughout his adult life. Milt was reliving the memories of his youth through the wide eyes of his son. The two would often fish the cool waters of the San Gabriel River. A sunlit clearing in the forest was always a good place for a family picnic, and the nearby perch offered Milt and Helen a chance to relive the romance of their beginnings in the mountain paradise. The pool at the newly renovated Mount Wilson Hotel provided some welcome relief from the heat on warm summer days (Figs. 5.1 and 5.2).

Fig. 5.1 Billy stands in front of the Humason's cottage on Mt. Wilson, 1917

Fig. 5.2 Winter wonderland:
Billy with the Humason's
dog, Trixie

Knowing nearly everyone working at or near the observatory helped to ease the transition as well. Having worked in almost every capacity from the base of the mountain to its peak Milt was very well known within the small community that lived and worked there. Not that Milt needed any help making friends, but his reputation as a former muleskinner who had married the daughter of the observatory's chief engineer was a good icebreaker when meeting new members of the staff. From the day he stepped foot onto the mountain in his new capacity as janitor, Milt began to win hearts and minds. A brilliant storyteller, horseman, poker player, fly-fisherman and practical joker, Milt endeared himself to the small but evolving band of scientists that occupied the observatory buildings and grounds.

As custodial positions were concerned working at a mountain observatory was not typical. Delirious astronomers, stumbling around in a dark room after a night of stargazing, tossed the developer used in exposing their photographic plates around like spaghetti sauce in an Italian kitchen. When cleaning up Milt often found tiny pools of the stuff dripping from the ceiling.

To guard the observatory's sensitive and expensive instruments, firebreaks had to be cleared and checked regularly during the hot and dry summer months. During the rainy season, the road through the observatory had to be shoveled after snowfalls, and help was sometimes needed clearing the road to and from the facility after landslides caused by frequent rains in the area. With the regular maintenance needed on the buildings and equipment at the observatory, there was always something that needed to be done. When he wasn't working in his capacity as janitor Milt often helped his father-in-law work on fitting the new buildings with the state-of-the-art electronics and switches he was installing. A new high-tension power system had been installed recently by the Southern California Edison Company to power the many instruments and buildings now nearing completion on the mountain.

The timing of his arrival back on Mount Wilson gave Milt another golden opportunity to get in on the ground level of the exciting new developments there. The increasing importance of stellar observations and the completion of the

100-inch reflector were a signal to George Hale that it was time to drop the word "solar" from the observatory's name. Writing in Volume 17 of the annual report to the Carnegie Institution, Hale states that the facility would henceforth be referred to as the "Mount Wilson Observatory." Although the Hooker Telescope was undergoing repairs to its mirror, mounting platform and observing platform, the telescope was thought to be nearly ready to come online. This meant that another position of night assistant would have to be filled to cover astronomers during observing runs on the new telescope.

Early on in his employment at the observatory, Milt was trained on the changing of the focus cages on the big reflectors. Although relatively new to the operation on Mount Wilson, the position of night assistant was highly valued. In time, the night assistants would become a respected and sometimes even revered group among the astronomers at the observatory. These skilled and knowledgeable technicians did everything necessary to set up the telescopes for their observing runs, clearing snow or debris from the domes and slits, positioning the telescopes according to the coordinates they were given by the astronomer, and working the controls when nature called and on food breaks. From their inception the night assistants were a united group and as necessary to the success of the stellar department as the setting Sun. Crossing one of these revered men meant trouble for any visiting astronomer no matter how well known or important his work was. To the night assistants, the mountain was their home, and their visibility and work ethic commanded respect from those who joined the ranks of the observatory.

The job of changing out the telescope cages involved a well synchronized and choreographed set of maneuvers to remove an existing cage and replace it with another. In spite of their great weight—40 tons for the 60-inch and 87 tons for the 100-inch—the big Mount Wilson reflectors were highly intricate instruments that required much care and maintenance to keep them sharply tuned for research. The primary mirror on the 100-inch telescope was ground so evenly that if you were to scale it to the size of the contiguous United States, there would be no variance in its surface greater than that of a basketball. From its perch on Mount Wilson the telescope could detect the light from a candle in New York City. Great care had to be taken to keep the instruments in good working order. The domes were kept dark day and night, and their outer domes were painted white to reflect as much of the Sun's heat as possible. Low wattage red light bulbs were used in place of brighter, warmer incandescent bulbs to reduce heat inside the dome. Shortly after sundown the dome slit was opened to allow the temperature inside to equalize with the outside air. The telescope cages could be changed out according to the focal length required by the astronomer for a given run.

Each of the reflectors had a set of three focus cages available for use in observing different objects. The short focus or Newtonian cage employed a flat mirror to reflect the desired image to an eyepiece near the top of the tube, while the longer Cassegrain focus collected the image near the base of the telescope mirror. Still longer, the image given by the Coudé focus was collected in a room beneath the telescope inside the dome building. Each setting required changing the telescope's cage. To do this, the telescope would have to be brought to the upright position using the motor control installed by Jerry Dowd. An assistant would climb onto the frame and loosen the bolts to release the cage. A crane attached to the dome would be driven along its rails to a position directly above the telescope, where a cable would then be lowered that would be attached by the night assistant or maintenance crew to the cage. Once secured the cage would be lifted from the telescope and lowered through an opening in the dome floor to a storage area under the observing platform. The crew then would attach the cable to the necessary cage for the observing run and the process would be reversed, as the new cage would be hoisted high into the dome above the telescope and then bolted in place on the end of the tube.

Showing the general aptitude and adaptability he had displayed all his life, Milt took to the work with his usual sense of purpose, curiosity and dependability. His lack of education seasoned an already humble spirit with an almost reverent respect that endeared him to many among the staff at Mount Wilson.

The patriotism and anti-German sentiments in the United States, due to the war raging in Europe, was boiling over onto the staff at the observatory in 1918. George Hale, who defined the mission of the observatory during the war as using "every available resource in aiding to overcome the enemy," had been stationed in Washington, D.C., since April 1916, working with the National Research Council to provide supervision and guidance on numerous subjects from the design of gun sights to bomb trajectory. In March of 1918, the observatory's chief instrument designer, Francis Pease, joined Hale in Washington to take the position of chief draftsman for the National Research Council.

By stepping up scientific research in defense of the country and its allies overseas, Hale hoped to advance the field of optics and other general fields in peacetime. To this end the shop in Pasadena was used in the fabrication and design of optics for gun sites and other weaponry. George Ritchey, having finished configuring the glass disc for the Hooker Telescope primary mirror, worked on optics for the Ordnance Department of the Army. Superintendent of the physical laboratory in Pasadena, Dr. Arthur King, and Dr. John Anderson worked on problems and designed, tested and constructed instruments for the Navy.

As he had done for several years leading up to the war, in his absence Hale had left Walter Adams, his steady right hand man and assistant director, in charge of the observatory. Adams, the son of American missionaries who was born in Syria on December 20, 1876, was a fair but frugal man who illustrated great constraint and budgetary discipline. On scientific matters, Adams could be brilliantly creative, but when it came to running an observatory he was strictly by the book and pinched his pennies.

As the fear and paranoia of enemy invasion of American soil increased the acting director had his hands full at the observatory. Many of those who were not actively serving wanted to help out in one way or another. The mood among the staff was vigilant and proud. The men were, not so quietly, forming a kind of observatory militia intent on keeping the facility safe against attack by German troops invading by sea along the Pacific Coast. Under continued pressure to authorize their actions, Adams made repeated appeals to Hale to allow the men to engage in regular drills. At last, Hale relented after repeated inquiries from Adams.

Guard posts were soon set up, and the men engaged in maneuvers and target practice. With his brother and cousin away in the service of their country, Milt joined his colleagues and countrymen in preparing themselves and the observatory in case of an attack.

* * * * * * * * * * * * * * * * * * **

Victor Hugo Benioff first came to Mount Wilson during summer vacation from nearby Pomona College in 1917. A native Californian, born to a Russian father and an American mother on September 14, 1899, Benioff was an avid musician who played the violin, cello and piano, and lived with his mother in the Claremont area very near the ranch Milt and Helen owned in Monrovia.

Benioff had come to Mount Wilson on an internship working with the solar department using the tower telescopes. When he discovered that the young man had talent, assistant director Walter Adams invited him to volunteer at the observatory again the following summer. This time, however, Adams put the aspiring scientist to work studying variations in velocity, spectrum and light for a group of Cepheid variables in Sagittarius. Benioff arrived on his summer break in 1918, and soon after he met Milton Humason. In spite of being eight years apart in age, their common interest in fishing and hiking and the closeness of their family homes in the valley gave them endless stories to share with one another. To Benioff, Milt was a survivor of the old world, cursing and chewing tobacco as he regaled him with stories of life in the early days on Mount Wilson. During the summer of 1918 the two formed a lasting friendship.

In his work on Cepheids for Adams, Benioff was studying star fields using the new 10-inch refracting telescope and camera that was housed in a small dome on the observatory grounds. Milt had been training on cage changes and the operation of the large reflectors. Working with the awe-inspiring telescopes was captivating to them both. Sensing Milt's interest in observing the stars, Benioff invited him out one night to try his hand at photographing them using the 10-inch telescope and camera. Milt eagerly accepted the offer, and the two met after dark on a clear night shortly after. During the observing run, Benioff quickly learned that his new friend was a natural for stellar research. Milt's color blindness gave him unusually acute night vision. Colorblind people (most of whom are men) very often possess the ability to see the contrast between black and white more sharply than normally sighted people. Milt was no exception and he took to the craft with ease.

Seeing his potential, Benioff suggested that they spend more time on it so that Milt could fully grasp the basics of stellar photography. By the end of that summer Milt was already making excellent images with the camera and telescope. Benioff was so impressed by Milt's progress he began singing his praises to Seth Nicholson, one of the observatory's new hires. Born Seth Barnes Nicholson on November 12, 1891, Nicholson grew up in rural Illinois and earned his undergraduate degree at Drake University. He was an extremely versatile scientist who worked in both the solar and stellar departments and was also serving as the chief draftsman in the drafting department at the facility. Having discovered Jupiter's ninth moon, Sinope, while studying the orbit of its eighth for his senior thesis at the Lick Observatory (immediately changing the subject of his thesis), Nicholson had become a brilliant astronomer and a master at determining orbits. So impressive was Nicholson's resume that George Hale had recruited him to work on his sunspot polarity program with the solar tower telescopes. The two had met shortly after Milt and his family moved to the mountain.

Again, Milt's country charm and knowledge of the mountain and its history impressed Nicholson. In turn, the measure of respect afforded him by this towering intellect was reassuring to Milt in the early days when he was taking his first tentative steps into the world of astronomy. The likeness in age between the two and their shared wry sense of humor soon made them close friends. Throughout his life at Mount Wilson, Seth Nicholson would remain Milt's closest friend. Naturally, when he heard Benioff rave about his buddy's newfound ability in stellar photography, Nicholson couldn't resist teaching Milt the basic mathematics used in astronomy. In the fall of 1918 he started tutoring Milt whenever opportunity struck.

On November 24, a weather event occurred on Mount Wilson that reminded everyone how severe conditions could get on the mountain. A storm brought with it sustained winds of 50 miles per hour, with gusts up to 90 miles per hour. The storm rattled Milt and family as they huddled in their tiny cottage on the hill. After 52 h the weather finally cleared, and Milt went out to survey the scene. To his astonishment none of the observatory buildings had sustained damage. A few pine trees had been uprooted near the Hooker Telescope dome and would have to be cleared, but, all things considered, the observatory had stood up well to the storm.

* * * * * * * * * * * * * * * * * **

The unofficial end of World War I, in November of 1918, brought with it the slow return of staff members and normal operations at the observatory. As Europe began picking up the pieces from the devastation and the tremendous loss of life, the world began to return slowly to peaceful prewar conditions. Around the observatory spirits were riding high on the news of the return of George Hale in December. The director had taken time out from his war work in Washington, D.C., to come to Pasadena for the marriage of his daughter, Margaret, to Paul Scherer, the son of Throop University president James Scherer. The young man had recently returned from war service himself, and the couple were married on Christmas Day.

One afternoon, a week or two later, Milt had a visit from Jerry Dowd while he was at work at the observatory. A mountain lion had made its way to the Dowd's ranch on the mountain and snatched one of his goats. Dowd had come to Milt for assistance in dealing with the problem. The lion's daring and stealth was not only an irritation to Dowd, who feared he would continue losing his livestock if something wasn't done, but could potentially be a threat to the safety of young Billy, who visited the ranch frequently with his mother. With new high-tension wires supplying electricity to the observatory by Southern California Edison, Dowd was busy installing the controls for the 100-inch telescope and 150-foot solar tower telescope, rewiring the equipment in the powerhouse and installing a new large switchboard and automatic switches for the laboratory at the observatories head-quarters on Santa Barbara Street in Pasadena. Besides, Milt's knowledge of the mountain and experience hunting and fishing made him better suited to the task than the 50-year-old Mount Wilson electrical engineer.

Apprehensive but resolute, Milt set out to trap and kill the lion. After some searching he found the trail of the cat and tracked it for some distance. When he felt he had got close enough for the lion to be lured by the scent of fresh meat he set a trap using some fresh kill he had made as bait. Milt carefully camouflaged the trap and skulked away quietly. The next day he returned to the site to check on the trap and found it empty. The sly cougar had snatched the bait from the trap's steel jaws and dragged it away to an easy meal. Milt stooped over the trap examining the jaws, wondering how the big cat had outmaneuvered him. The next moment he looked up and was surprised to find himself face to face with the mountain lion. The hunter had become the hunted. In the seconds that followed the two faced each other, the lion crouching low, its steely eyes fixed on Milt whose heart was thumping in his throat. After what seemed an eternity the lion bared its teeth and lunged at Milt who instinctively raised his rifle and fired, stopping the lion in its tracks. Standing over the fallen beast in the stillness of the cool winter day, his body buzzing with adrenaline as the echo from his rifle shot slowly fading into the wind, was as solemn an experience as it was electrifying for Milt. He didn't hunt for hunting sake and was more interested in maintaining the balance of nature than trophies for his mantle, but the press had other ideas. A photograph taken for a local newspaper commemorates the occasion. In it Milt stands in an opening in the forest, his rifle by his side, wearing a cowboy hat with the cat draped around his neck. At the observatory, word of Milt's harrowing adventure brought him even greater fame among the staff. For years after the incident, when asked by a visitor to the mountain whether there were lions still about, Seth Nicholson would wink at them and say, "Well, there are some, but Milt's got most of them." Wendell P. Hoge, a night-time assistant and friend summed up the event in a poem he entered into the observatory logbook on Sunday, January 19, 1919:

A mountain lion got Dowd's goat,
So Milton went a-hunting,
And now he has a lion skin,
To wrap up Baby Bunting.

The lion was a mountain beast,
More than a hundred weight,
From tip of nose to end of tail,
She measured five foot eight.

Astronomers are very brave,
To work up there all night,
With lions roaring all around,
Most folks would faint with fright.

They're brave all right, but then we're sure,
They'll all be glad to note,
No more will roam around the dome,
The beast that got Dowd's goat.

* * * * * * * * * * * * * * * * **

George Hale, acting chairman of the board at the National Research Council during the war, resigned his position and returned to Pasadena with Francis Pease in April. Around that time the physical laboratory and optical shops stopped work on projectiles and optics and began devoting their full time again to scientific problems. Work was finished on the 100-inch telescope, and tests were being run to determine its true capabilities. As the new telescope was being prepared to come online, Hale sensed that a paradigm shift would need to be made for the observatory. In a running narrative in the Contributions of the Carnegie Institution Yearbook from 1919, Hale laid out this new vision in broad detail, starting with the observatory's original mission: "The purpose of the Observatory, as defined at the time of its inception, was to undertake a general investigation of stellar evolution, laying special emphasis upon the study of the sun…"

Since then, Hale said, developments in both instrumentation and methods of study had extended the means through which progress could be made on celestial problems such as the structure of the sidereal universe, which were not regarded as being part of the original plan. The recent updates to the equipment at the observatory and its participation in Professor Jacobus Kapteyn's international study to determine galactic structure, Hale continued, had "naturally and inevitably led to a very considerable extension of our original program." Ultimately, Hale believed, the varied approaches would be indelibly tied together as investigations of sidereal problems led to greater understanding of stellar evolution and vice versa. As the roster of scientists grew at the observatory, Hale was free to once again pursue his favorite subject, the evolution of the Sun.

* * * * * * * * * * * * * * * * **

The signing of the Treaty of Versailles on June 28, 1919, effectively ended World War I and created the League of Nations. But public mistrust of the league fomented by political opposition in Congress kept the United States from ratifying the measure. President Woodrow Wilson, whose earlier stellar reputation was badly damaged by U.S. involvement in the war and the creation of the league intended to stop any such conflict happening again in the future, was preparing a whistlestop tour of the country in support of it. The decimation and alteration of the landscape of Europe was considerable. The German, Austro-Hungarian, Ottoman and Russian empires were swept away in waves of revolution that ensued near the end of the war. In the carnage, France, Germany and Russia had lost millions of soldiers and civilians, accounting for 10 to 20 percent of their active male populations. Once great cities from Belgium to Turkey lay in ruins, leaving millions without food and shelter. Cases of typhus and malaria were increasing, and an epidemic of Spanish flu had gripped the world in 1918, killing millions more. Although America's involvement in the war had been limited in both time and scope, the loss of tens of thousands of American soldiers had a profound and lasting effect on the public sentiment toward armed conflict.

* * * * * * * * * * * * * * * * * **

Hugo Benioff returned during his summer break in 1919 to assist in both stellar and solar research. On a visit to the Humason's cottage Milt filled in his friend on the developments of the year, his work with Seth Nicholson and his run in with the mountain lion. Billy was enjoying the mountain as only a 6-year-old boy could, and Milt and Helen had decided to sell their ranch in the valley and settle into life at the observatory for a while.

Nicholson had been spending quite a bit of time getting Milt acquainted with the basics of astronomy while Milt continued to hone his observing skills at the telescope. On August 21, 1919, the Reverend Joel H. Metcalf, a Unitarian minister at Camp Idlewild, Vermont, found a comet while making a sky survey. The comet had first been discovered in 1847 by a man named Theodor Brorsen in Germany, and several people at different observatories around the world were now studying it to confirm whether it was Brorsen's comet or a new comet of Metcalf's discovering. Nicholson decided that he and Humason should work together on a report of the comet for submission to the Proceedings of the Astronomical Society of the Pacific. On September 15 Milt went out for a night's observing at the 10-inch telescope, carefully guiding the telescope across the sky on the warm summer night. Seeing on Mount Wilson is best during the summer months, and Milt didn't have much trouble in getting a good plate of his target. At 7:14 he began a one-hour exposure that should have been enough to do the job. He removed the plate from the telescope and headed to the dark room at the lab to develop it. The next day Nicholson walked Milt through an analysis of the comet, showing him how to measure its shape, and the two discussed which of its attributes would be interesting to include in their report. As stated in the paragraph-length report, the exposure revealed "the comet as a round nebulosity about 12 min of arc in diameter with a faint tail, three and a half or four degrees long." The comet, the report continued, "showed no stellar nucleus," and its narrow tail "spread but little as it receded" from the head, "reaching a width of about 6 min of arc at a distance of two and half degrees." With the publication of this short study, submitted to the PASP in 1919 under the title "Metcalf's First Comet," Milton Humason had made his first small report on existing conditions within the Solar System.

* * * * * * * * * * * * * * * * * **

On October 2, 1919, President Wilson suffered a massive stroke that nearly killed him and left him entirely paralyzed on his left side. Although he survived the ordeal, his condition left Wilson powerless to pursue the ratification of the Treaty of

Versailles, and his coveted League of Nations, in Congress. No longer able to serve in his capacity as president, Wilson would spend the remainder of his term under the care of his wife and doctor. Congress tried unsuccessfully to ratify the treaty several times until March of 1920. In November, Republican candidate Warren G. Harding would be elected president in a landslide victory. Then chairman of the Senate Foreign Relations Committee, Henry Cabot Lodge, one of Woodrow Wilson's arch rivals and an outspoken critic of the League of Nations, pronounced the league dead as far as the United States was concerned. In a small measure of redemption, Wilson would be awarded the Nobel Peace Prize a few weeks later.

* * * * * * * * * * * * * * * * * **

By September, the 100-inch telescope was in regular use and undergoing extensive tests of its light-gathering capabilities. Dowd was finishing up the wiring of the telescope and dome, and a new cottage was being built by Mr. Jones and his team to house the night assistant for the new telescope.

On September 12, two forest fires broke out about 20 miles from the observatory on opposite sides of the mountain. The terribly destructive fires raged out of control for nearly two weeks, threatening the observatory and grounds. Observing was entirely stopped, as the fires created huge plumes of smoke that filled the air. In an effort to prevent the fires from reaching the domes and other buildings, fire breaks were expanded, and the fire system was held at the ready. Fortunately, a heavy storm rolled over the area on September 26, dumping torrents of rain and extinguishing the fires.

* * * * * * * * * * * * * * * * * **

The news of his first publication with Seth Nicholson heightened Milt's reputation among the observatory staff. Stories began to circulate of the high school dropout who had gone from muleskinner to rancher to janitor to part-time night assistant and was now had recently become a published observer.

One of those looking on with great interest in Milt's education was Harlow Shapley, who had heard of his exploits through Hugo Benioff, whom Shapley had mentored over the summer. Shapley and Benioff were working on a new means of photographing "very faint stars" at the 134-foot focus of the Hooker Telescope. Milt's growing abilities and willingness to immerse himself in the field intrigued Shapley, who had himself found his way to the science by unfocused means. The native of Missouri had been a staff reporter for the local newspaper in Joplin when he was accepted to the University of Missouri at Columbia, where he graduated with a B.A. degree in 1910 at the age of 25. He then returned to the university hoping to earn a graduate degree in journalism, but no such major existed at the time.

Not sure what he should do next, Shapley decided to start at the beginning of the list of courses offered and work his way down until he found something he liked. His first choice was archaeology, but the course was full, so he scrolled down the list a bit further and found astronomy. Shapley enrolled in the course and received a Master's degree in astronomy at the university in 1911. Shapley had joined

Frederick Seares in the stellar photometry department at the Mount Wilson Observatory in 1914, and a mere four years later had culled his research on globular clusters into a seminal work that included a conclusive determination of the size and shape of the Milky Way Galaxy and the eccentric position of the Sun and Solar System within it.

On the heels of the publication of several monumental papers on galactic structure, Shapley was becoming a rising star in the field of astronomy and on the campus of the Mount Wilson Observatory. An educator at heart Shapley sought Milt out and began introducing him to the science of photometry. As part of his study, Shapley employed his new apprentice in blinking plates he had made on the stereo comparator, looking for signs of Cepheid variables. It was during this process that Milt made his first discovery of a nova in Andromeda. It was the seventeenth nova found in what was then referred to as the Andromeda Nebula, and would make a perfect subject for Milt's second report to the PASP. The report, entitled "A Seventeenth Nova in the Andromeda Nebula," briefly detailed the location of the nova, "15° east, 150° north of the nucleus, and surrounded by soft nebulosity," on plates made by Shapley and Roscoe Sanford on the 60-inch telescope.

The hook was in. Milt was consumed with the field of astronomy and was doing everything he could to further immerse himself in the work. He sought every opportunity to work on astronomical problems that were suitable for him to work on, with his limited experience and knowledge. One of these problems was determining the presence and location of the mysterious Planet X, a supposed ninth planet thought to be orbiting the distant reaches of the Solar System. More careful naming of the planet along linear numeric lines might have led to it being named Planet IX, but we won't split hairs. Percival Lowell, founder of the Lowell Observatory in Flagstaff, Arizona, had theorized that wobbles in the orbits of Neptune and Uranus were caused by the presence of an as-yet undiscovered planet. Lowell had searched in vain for the planet until his death in 1916.

Late in December of 1919, Milt made his own unsuccessful attempt to locate Pluto among the dark patches of the Moonless night. Although the search ended without result, the project had provided a useful target for Milt to hone his skills as an observer and a stellar photographer. The distance and very low brightness of the object he was trying to capture meant very careful and patient scanning of the night sky with a steady hand to produce the clearest possible negative plates. The planet's small size and distance would have made it difficult for a fledgling observer using a telescope of such small aperture to find in the sea of distant light, even following its assumed orbit. W.H. Pickering, an experienced astronomer, couldn't locate Pluto using the 24-inch Clark refractor at Lowell Observatory during his own search. Had Milt tried again in the mid-1920s once he had gained more experience he might have been successful, but by then he was onto other matters.

The end of the war brought Lewis and cousin David Witmer home to Los Angeles and, with their heroes safely returned, the Christmas of 1919 would be a year to remember. Lewis was working as a repairman at a local phonograph company, having moved back in with his parents and sister at the family's new home at 323 Witmer Street. The house had been designed by David Witmer. It was one of several homes the aspiring architect had designed in the area. He and his young wife, Helen, were planning to move into a home of their own at 210 Witmer, and David's mother and sister planned to live next door at 208 Witmer. Virginia, who would be celebrating her 21st birthday soon, was living at home with her parents and working as an assistant manager at an insurance company.

As the holiday approached Milt and Billy went for a walk to find a tree they could cut down to display in the family cottage. After some hiking and conversation, Billy finally settled on a large pine tree a half mile or so from home. The tree was so tall Milt had to cut it off 5 feet above the ground, and even then it barely fit through the door. That year on Christmas Eve, Milt returned from the observatory after a night of assisting at the 60-inch telescope. The house was quiet and the tree stood near the chimney of the tiny cottage, beautifully trimmed and brimming with gifts that Helen had wrapped and placed under it. Ever the enthusiastic play actor, Milt had an idea. Quietly walking around to the side of the house where the slope of the mountain brought him closer to the eaves, Milt climbed onto the roof carrying with him a set of sleigh bells and started stomping around jingling the bells and pretending to be jolly old Saint Nick come to deliver gifts. Billy sat up in bed when he heard the rumpus on the rooftop. There were footsteps, some shuffling and then a shushing sound. Milt had slid off the roof and was making his way back around to the front door of the cottage where he entered and continued his merry deception. Billy could barely contain himself when he heard Santa's footsteps inside the house but decided not to disturb him so as not to scare him away. On Christmas morning, Billy ran out to the living room to find a new train set with a steam engine that chugged around the tracks billowing tiny puffs of smoke out of its boiler stack. Later that day the family went sledding on Jones Hill behind the 100-inch dome.

* * * * * * * * * * * * * * * * * **

As the New Year began the American public was bracing itself for life without alcohol, or at least the appearance of that life. The Eighteenth Amendment, which made the production, transport and sale of alcohol illegal in the United States, was set to take effect on January 17, 1920. The new law and accompanying Volstead Act had been ratified in January of 1919 and signed into law over a presidential veto. The result of a social battle between religious conservatives and other elements of the shrinking rural United States and more progressive people in the ever-growing urban areas, most believed the new law would simply cause those who wanted to continue producing, selling and drinking alcoholic beverages to do it inconspicuously, which they did.

Milton Humason was no exception. Even before the turn of the New Year, Milt had coordinated his local ranching network in an effort to supply him with a steady supply of booze. Soon he began brandishing the homegrown substance at poker games and around the billiards table. Throughout the Prohibition Era Milt carried around a flask of the rancid liquor that he liked to call panther pacifier (Never a heavy drinker, Milt preferred Jack Daniels once Prohibition ended). One taste would send the average man reeling in anguish, the fiery sting of the potion burning his mouth and throat. In the years to come, Milt's mountain moonshine would become legendary, boosting his already cult-like status among the staff at the observatory as a true man of the mountain.

* * * * * * * * * * * * * * * * * * **

Despite his setback in the Pluto experiment, Milt's ambition, curiosity and increasing facility were paying off for him. Harlow Shapley, impressed by his progress, sang Milt's praises to the assistant director, Walter Adams. Having heard similar praise from Hugo Benioff and Seth Nicholson, Adams decided to give Milt a larger chunk of the observing time assisting the observatory staff. In a census report taken February 4, 1920 on Mount Wilson, Milt lists his occupation as "Assistant Observer." As exciting as it was the new position meant spending more time working nights. Milt and Helen would have to adjust to a new schedule to allow 7-year-old Billy time with his dad. On observing nights Milt left the family cottage in the afternoon and didn't usually return until just before dawn, when he would go to sleep and wake up around midday. Despite missing her husband's presence at home on those nights Milt was at the telescope, Helen occupied her time with frequent visits to her mother at the Dowd's ranch. On nights when the weather was good, Milt could count on a visit from Helen and Billy with snacks for him and his work mates at the domes where they worked.

As part of his continuing education Adams decided to familiarize Milt with his technique of determining stellar parallax using spectroscopy. As Hale stated in his remarks to the Carnegie Institution, Adams's new "means of attack…through the method of spectrographic parallaxes" had led to a "considerable extension" of the research at the observatory. Adams' inventiveness had created a shortcut for determining stellar distances that vastly increased the efficiency of collecting useful data on the stars. In the years before Adams discovered this new method, photometric measures of stellar parallax were done with painstakingly minute detail and resulted in low numbers of stars for which useful data was collected. Using the new method the stellar department was yielding measurements of hundreds of stars per year. Adams would be awarded the Bruce Medal in 1928 for this advance.

Milt's energy and enthusiasm for the work was boundless in the early days of his development as an observer. The sense of purpose and the sheer desire to get back to the telescope were driving him toward the mastery of his new craft at an amazingly fast rate. In May Milt photographed the spectra of the recurrent nova, T-Pyxidis, with the spectrograph attached to the 10-inch telescope for his third paper, which was published in the *PASP* in 1920. By August, as part of his work for Shapley and the photometry department, he had made 46 one-hour exposures of "star-fields." The ease with which he picked up the techniques he was learning and his facility and care with the instruments was winning him much appreciation among the staff, who increasingly sought him out for collaboration. Walter Adams was one of those who saw Milt's potential. Adams became Milt's greatest champion in those days, suggesting to George Hale that Milt be appointed to a full-time position on the staff. But despite Adams' recommendation Hale believed Milt to be too old and too uneducated for a role on the staff at Mount Wilson. As part of the mandate of the Carnegie Institution, Hale preferred to utilize the cash made available by the Carnegie Trust to bring in more seasoned and better educated scientists. After all, he had two of the biggest telescopes in the world sitting on top of Mount Wilson. Why waste his time training a nearly 30-year-old ex-muleskinner when he was attracting top talent to work at the facility. Conceding the point, Adams let the issue drop for the moment.

Although he was disappointed at the news, Milt contented himself with working at his new job by night and spending time with his family by day. Regardless of the outcome regarding his appointment to the observatory, his transition back to Mount Wilson was complete. He was no longer the wide-eyed boy who had first hiked the mountain trail in 1903, but he retained the same wide-eyed spirit, amiable nature and charisma so characteristic of his personality throughout his life.

Chapter 6
The Spectroscopists

Abstract Milt continues to publish reports on cosmic events while biding his time as an assistant to the sidereal staff. As a night-time assistant he has a run in with a particularly well known scientist, Henry Norris Russell who is a regular visitor to the observatory. An event surrounding the repair of the 100-inch telescope seals Milt's promotion to the staff at Mount Wilson.

Let's begin this chapter by paying homage to the computer, not computers as we know them today but how they were known in Milton Humason's era, when they were commonly referred to as either Miss or Mrs. From the early days of the Harvard Observatory's ambitious campaign to chart and classify the stars, which eventually resulted in the *Henry Draper Catalogue*, women had been employed to reduce the data from photographic plates. As more institutions were created in the latter twentieth and early twenty-first centuries, female "computers" were employed to convert the increasing amounts of data that poured in from astronomers using an ever-evolving array of instruments and formulas. These women were the unsung heroes of astronomy well into the 1950s, and a few of them were responsible for breakthroughs that helped build the foundation of modern cosmology.

Three of these women—Annie Jump Cannon, Williamina Fleming and Antonia Maury—helped Harvard Observatory director Edward C. Pickering develop the stellar spectral classification system for the *Draper Catalogue*, which is still used as the basis for charting stars to this day. Another, and maybe the best known of all the original group of female computers at Harvard, Henrietta Swan Leavitt, figures prominently later in the story. None of the developments in astronomy during this period happened without the steadfast dedication, creativity and breathtaking insight of these remarkable women. Always mentioned lastly among the staff in the Mount Wilson, it is only fitting that they be given first light in the pages of this chapter. One can only wonder what the state of astronomy might be today if women had been given a greater presence during the early days of this revolutionary period in the field.

* * * * * * * * * * * * * * * * * *

Through the kindness of friends and colleagues, Milt began to learn not only the science but some of the history surrounding his new job. The science of spectroscopy

© Springer Science+Business Media New York 2016
R.L. Voller, *The Muleskinner and the Stars*,
Springer Biographies, DOI 10.1007/978-1-4939-2880-4_6

had its roots with Sir Isaac Newton, who first applied the word "spectrum" to the rainbow of colors made by white light when it is passed through a prism. More than a hundred years later, Joseph von Fraunhofer began experimenting with the first spectrometers, and the science was pushed forward through the mid-1800s by a legion of scientific stalwarts, from William Herschel and Léon Foucalt to Anders Angstrom to William Thomson (Lord Kelvin) and Robert Bunsen, all working on problems concerning the spectra of gases produced by the ionization of different elements in stellar atmospheres. The image of George Hale's mentor, William Huggins, hanging out in the library at the observatory offices in Pasadena, was illustrative of the contribution of spectroscopy to the understanding of stellar evolution and universal structure. In the 1860s, Huggins and his wife Margaret began measuring the stars using spectroscopy and determined that they were made up of some of the same elements as Earth. In 1885, Johann Balmer accurately predicted the appearance of a series of hydrogen absorption lines in the spectra of stars. Four of these are located in the visible part of the spectrum. In the early 1920s spectroscopists were using the spectra of stars (star clusters and nebulae as well) to determine things such as surface temperature, which was a main ingredient in their classification.

As Milt began his career as a night-time assistant, the sidereal department at the observatory was ascendant. Of the two departments on Mount Wilson the night-time crowd were starting to get most of the press. Observing the stars through the lens of the 60-inch telescope was open to the public on Friday nights, and the 100-inch telescope was attracting 10,000 visitors per year (Fig. 6.1).

Although long term programs such as Hale's research on the Sun's cycle were in progress and showed future promise in understanding the life of a star, the solar department hadn't had a significant headline-making breakthrough since Hale's discovery of the magnetic fields in sunspots in 1908. More significant work was not being done at Mount Wilson than in the solar department, but the pace of the work and the dense scientific nature of the data was not a good source of news for the general public. As far as most people were concerned the Sun, a star though it might be, just wasn't that exciting a topic.

Meanwhile Harlow Shapley's recent prediction of the size, shape and makeup of the Milky Way was grabbing headlines and causing controversy in both the scientific and public spheres. His 1920 public disagreement with Hebert Curtis on whether all matter existed within one enormous all-encompassing universe (as Shapley believed) or many "island universes" was one of the most talked about events in science. After dueling to a draw with Curtis in their debate in Washington, D.C., Shapley was preparing to leave Mount Wilson in 1921 to assume the directorship at Harvard Observatory, left open by the death of Edward C. Pickering in 1919.

Due to the work of Shapley and others in the field, such as Jacobus Kapteyn, Hale had begun to alter the direction of the observatory to allow for broader research of sidereal problems. The joint program established by Kapteyn to map 206 areas of the sky, measuring the proper motions and radial velocities of stars using photometric and spectroscopic methods for the determination of galactic structure, was well underway. The general theory of relativity was garnering

Fig. 6.1 On a fishing trip to the cottage at the West Fork, Milt sits smoking a pipe watching his mother and father, Laura and William Humason, at *left*. Billy sits at the foot of the cabin looking on

attention in all departments of the observatory by 1920. The implications to scientific theory caused by Albert Einstein's 1915 companion to his theory on special relativity (published in 1905) were being tested by research institutions around the world. At Mount Wilson, Albert Michelson was attempting to measure the displacement of a star caused by the gravity of Jupiter using an interferometer he had designed for use with the 100-inch telescope. Michelson was also working on an experiment to measure the velocity of light, getting help in this capacity from Francis Pease and several members of the optical and instrument departments. The exploration and experimentation created by the work of scientists such as Einstein, Kapteyn and Shapley, held possibilities for new discovery that propelled both stellar and solar work at the observatory. As time went on and new discoveries were made the mutual importance of research in all departments was becoming more and more evident.

Adventures in the Kapteyn Selected Areas and other programs reached nearly every department of sidereal research at Mount Wilson starting in 1909, spurring growth in both personnel and equipment. By 1920 the staff of spectroscopists, led by Walter Adams, included the names of men who would become titans of the field such as Frederick Seares, Alfred Joy, Paul Merrill and Edwin Hubble. Although the first and last name of the men who worked at the observatory was used in public, on the mountain everyone went by his last name.

In his capacity as night assistant, Milton Humason worked with all of these men starting in the fall of 1919. The demanding work of observing at the various focus cages of the big reflectors on Mount Wilson required a great deal of preparation, attention to detail, aerial daring in near total darkness, and copious amounts of stamina and patience. In preparing for an observing run an astronomer was to have both the right ascension and declination coordinates for the focus star and the object under study. Failure to have these two coordinates well defined upon arrival was considered to be a needless waste of the night assistant's time, and they were watching. The first law on the mountain was that never, under any circumstances, was it a good idea to ruffle the feathers of a night assistant. These capable and diligent observers were experts in the set up and use of their instruments and had the ear of nearly every member of the observatory staff. Slighting one, making him do the astronomer's work, making him feel unworthy, or somehow inferior, could make life difficult on an unsuspecting astronomer.

The noted dean of American astronomers, Henry Norris Russell, found this out the hard way. Starting in 1921, Russell, a monumental figure in astronomy during the first half of the twentieth century, worked as a research associate at Mount Wilson on his winter breaks from the Princeton University observatory, where he served as director. As part of his research Russell was studying the H and K lines of calcium in solar, stellar and laboratory spectra. These lines appeared much brighter than those of lighter elements such as sodium and magnesium. Russell's theory was that the enhanced brightness in the lines of calcium was due to the loss of one or more calcium ions in the solar atmosphere.

Milt, by then a seasoned assistant, quietly kept score, gauging the character as well as the ability and preparation of each man he came across. Russell put him off from the start, arrogantly commanding Milt to set the telescope in position and ordering him to have his eggs and toast ready for him when he arrived at the midnight lunch shack at 12 a.m. sharp! Milt quietly went about his business, deftly setting the telescope in position for Russell as he began his night's observing. As the hour approached for their break, Milt left his position to head to the lunch shack for a quick bite and to prepare Russell's eggs and toast. As he chatted away with his friend Wendell Hoge, who was assisting on the 60-inch telescope that evening, Milt watched as the egg came to a boil. When it had finished Milt removed the egg from the boiling water and then poured the steaming egg-water into Russell's teacup, setting it on the table beside the boiled egg and toast. Returning to the dome, Milt dutifully assumed the observing position at the telescope to allow Russell to head to the shack for his egg and toast. In a meeting with Walter Adams the next day, Russell extolled the virtues of the telescopes and other equipment at the observatory but complained vehemently about the extremely poor quality of the drinking water on the mountain.

With the publication of his third paper, Humason had made his first leap from direct stellar photography to spectroscopy. Noting this, Paul Merrill became the first of the spectroscopists to call on Milt for his skill at the telescopes. Merrill had joined the Mount Wilson observatory staff in 1919, at age 33, already established as a serious observational astronomer. After getting his Ph.D. in astronomy from UC Berkeley, where he worked on measuring standard radial velocity systems of wavelengths, he taught astronomy at the University of Michigan for a few years before transferring to Washington, D.C., to work at the Bureau of Standards. While at the bureau, Merrill advanced plate emulsion technology, dying plates with dicyanin to make them more sensitive to near-infrared wavelengths. This improvement would come in handy in his search for and study of long-period variables at Mount Wilson. One of Mount Wilson's masters of minutia, Merrill would be a leading contributor in solving some of the riddles of stellar and nebular evolution. His discovery of technetium in the spectrum of the S-type star, R-Andromeda, led to the development of the theory that nucleosynthesis exists within stars, an important step in the understanding of how planets were formed.

Merrill began his study of long-period variable stars immediately after his arrival in Pasadena. An arch conservative who wasn't given to suffering fools politely, Merrill began asking around for capable assistance and was led by Walter Adams and Harlow Shapley to enlist Humason for the project in 1920. Each would observe stars selected for the program taken from the *Draper Catalogue*. Merrill would make exposures of spectra using a spectrograph attached to both the 60- and 100-inch telescopes while Milt would do the same using the 10-inch refractor. The two published their first paper in the fall of 1920 on ten B-class stars with magnitudes ranging from 4.2 to 8.2. To ensure their measurements were in good agreement, spectrograms of several stars were taken with both the 10-inch and one of the large reflectors. One feature these stars had in common was that the H-alpha lines (the hydrogen line in the Balmer series nearest the red) in their spectra were bright, indicating they had high surface temperatures. Many of these giant stars can be seen with the naked eye. Among the most prominent is Regulus in the constellation Leo, one of the brightest stars in the northern hemisphere.

Impressed by Humason's ability, Merrill asked Adams to let his apprentice take some observing time on the 100-inch telescope, which would free Merrill to work on photometric measurements and other details. Adams agreed, and starting in December of that year, Milt began observing runs on the 100-inch telescope, sharing the load almost equally with Merrill on the 100-inch reflector. In all Milt accounted for eighteen of the twenty-five plates taken for use in their second paper compared with fifteen out of the thirty-five total plates from their first.

Merrill and Humason published two new lists of class B stars in 1921. In the third list Milt accounted for twenty-four out of the twenty-seven plates used in their study. By now Merrill had full faith in his ability with the large reflectors for both direct photography and spectroscopy.

The high praise from Merrill as to Milt's increasing facility convinced Adams that he might be of use to Frederick Seares, who was conducting a broad survey of photographic and photo-visual magnitudes of stars near the North Pole. At the time Seares, who had been born May 17, 1878, in Cassopolis, Michigan, was serving as chairman of the International Committee on Magnitudes and senior member of the photometry department at Mount Wilson. With Shapley's departure in March of that year, he was the only member of that department until Walter Baade came to Mount Wilson in 1931. Seares had come to the observatory in 1909 after serving as director of the Laws Observatory at the University of Missouri, where Shapley had earned his undergraduate degree in astronomy. The sheer scope and detail of his work (1909–1940) would make Seares the dean of astronomical photometry worldwide. During his career he would publish over 180 papers on stellar magnitudes and star distribution throughout the Milky Way Galaxy and serve as president of the IAU's commission on stellar photometry for fifteen years. When he wasn't conducting his meticulous study of the stars, Seares served as superintendent of the computing division, was in charge of increasing and organizing the library's inventory and was editor of the Mount Wilson Contributions to the Carnegie Yearbook.

To crosscheck his magnitude scales of northern polar stars taken with the 60-inch reflector, Seares asked Milt to measure them using the 10-inch Cooke refractor. The telescope, with its relatively small aperture and focal length as related to the 60-inch reflector, would give good results on the visual and photo-visual scale of stars in the region fainter than the eighth magnitude. During their research, Seares showed Milt how to measure the stars using photometry. Milt photographed and measured numerous stars between the fourth and fourteenth magnitudes for the study, handing off his data to Mary Joyner in the computing department for reduction before Seares analyzed them. As 1922 rolled around the team was preparing to publish a long paper on the subject of brightness among stars in the north celestial polar region. Humason's work confirmed Seares' earlier scale measurements from his 1915 paper on the subject. Although he was as meticulous in the details of operating a telescope as Seares was in reducing the data from them, Milt apparently didn't care much for that kind of work, preferring instead to spend his nights scouring the sky at the controls of the giant reflectors.

* * * * * * * * * * * * * * * * * **

In March of 1921 the Humason clan had a much anticipated visitor to Los Angeles. While attending boarding school in England in 1915, Virginia had met and befriended a Dutch girl. The girl was a member of local tennis club and the friend of Billy Suermondt, a law student at Leiden who was preparing for a trip to the United States on the advice of his father. As he prepared for his voyage Suermondt ran into the girl who told him, "If you go to California, please go to Pasadena and give my love to Virginia Humason." Billy packed his things and set out for the United States on October 31, 1916.

Around the New Year young Billy Suermondt stopped by the Humason home to bring greetings to Ginny from her Dutch friend in Holland. Sparks flew between them, and they fell for each other instantly. After some weeks spending time together it was time for Billy to leave, but the die was cast. The completion of his law degree and two years of service for his country ensued, but Billy and Ginny never lost track of each other. Finally, after years of waiting, they were united again. This time, Billy had no intention of leaving his beloved Virginia behind. The young couple soon boarded a ship bound for Dutch shores and they were married in Rotterdam on October 25, 1921.

It was a harsh winter. Snowfall for the year 1921–22 was the highest since records had been kept at the observatory. During one storm in December nearly 30 inches of snow fell, sealing off access to the mountain for days. As the sky cleared, walkways were shoveled out for passage from one building to the next, but the domes remained covered in snow. Snow falling from the dome slit could damage the telescope and instrumentation, so it had to be removed from the top of the dome. This assignment was given to the night assistant on duty, and Milt was the assistant scheduled on the 100-inch that week. Tied to the top of the dome over 100 ft. above the ground, his nerves standing on end, Milt carefully shoveled the snow from the slit doors, making his way slowly down the doors until all the snow was removed.

Milt survived the ordeal, but the snowy winter and conditions on the mountain were starting to raise a bigger question for him and Helen. Billy was getting older, and his schooling was becoming a key priority for them. Milt's salary as a night assistant wasn't quite good enough to afford a home in the valley, and George Hale didn't seem interested in promoting him to the staff of astronomers. Although they loved the mountain Milt and Helen knew they were going to have to find a way to move back to the city soon.

Now entering his third year of research in the field, Humason was gaining a reputation for observing and photographing faint objects and getting good, usable results. As a consequence he found himself increasingly busy as one astronomer after another was asking him to lend his expertise to his work. In all, Milt published eight papers that year, four of them on Merrill's B-class star program of long-period variables. While working with the 10-inch telescope on June 21, Milt discovered a new planetary nebula that became the focus of one his papers. On a plate he made of Andromeda two days later with the 100-inch telescope, Milt discovered a new nova. Gaining confidence from others who were depending ever more on the sharpness of his eye, Milt was certainly starting to feel very comfortable in his role as an observer and spectroscopist. The satisfaction he felt in his work was countered, though, by the very real problem of needing to somehow get off the mountain on a permanent basis. Although he was getting good results at the telescopes the news had still not swayed Hale to promote him (Fig. 6.2).

In January a problem with the 100-inch dome's rotation system threatened to shut down operation of the telescope for six weeks. Inadequate padding on the trucks that carried the dome was causing it to shudder, creating tremors in the images made at the telescope. To fix the problem the entire dome would have to be jacked up, and Adams was forced to shut down operation of the telescope while the repairs were being made. Humason wandered across the footbridge to the dome where Adams, Jerry Dowd and observatory construction head, George Jones, were standing. Adams related the problem to Milt, who looked up at the dome gleaming in the cool winter Sun. As he stood listening to the conversation an idea occurred to him. He turned and asked Adams if the slit doors and the telescope itself would still be operational while the tracking system was being repaired. Adams saw no reason why they wouldn't be. In that case, Milt suggested, why not set the dome slit pointing directly east before it was jacked up for repairs. This way stars rising in the sky within the dome opening could be studied while the truck system was being worked on.

Adam's penchant for frugality in the operation of the observatory was legendary. He insisted that no light bulb stronger than 25 W be used in the domes, not to keep

Fig. 6.2 Milt pauses from shoveling near the monastery after a snowstorm on Mt. Wilson, during the winter of 1921–22

the operating temperature inside the observing platform down but to save on the electric bill. Visitors to the midnight lunch shack during observing runs were treated to one hard-boiled egg, one piece of toast and tea. Any idea that kept the observatory running smoothly and efficiently was well received, and this was a brilliant idea. The assistant director looked sideways at Dowd and then returned his gaze to the ex-muleskinner turned fledgling astronomer with a smile on his face. He instructed Jones to have the dome set according to Humason's recommendation. Turning back to Humason, Adams informed him that he would be scheduling regularly during the period when the dome was under maintenance to work on Adam's and Joy's program charting and classifying stars in the Kapteyn Selected Areas.

There was no question to anyone who knew them that Milt was endearing himself to the assistant director. In late May, George Hale was preparing to leave the observatory for a year abroad to cope with the recurrence of the mental illness that had plagued him for many years. His work for the National Research Council during the war had served to heal his condition temporarily, but after two years at the controls of the observatory his health was once again in decline. At just 52 the director was contemplating stepping away from the daily operations of the observatory he so dearly loved.

As the two discussed the course of action at the observatory the subject of adding to the staff came up, and Adams once again threw the full weight of his support behind Milt's promotion. Citing Milt's ingenuity in setting up the 100-inch telescope while the dome truck system was being repaired as well as the night assistant's work with Shapley, Seares, Merrill, and Joy, Adams was more certain than ever that the move would pay off. This time Hale listened more intently to his assistant director. After years of dutiful and loyal partnership in which Adams had repeatedly showed himself a responsible and intelligent leader, Hale trusted him with nearly every aspect of the observatory's operation. Hale conferred with Adams for a few minutes and eventually came around to Adams' point of view.

That fall Milton Humason was promoted to the observatory staff at Mount Wilson. The promotion included a salary increase that would allow him and Helen to move the family down the mountain, where Billy could continue his education in the Los Angeles public schools. Wendell Hoge, one of Milt's close friends, was among those who had witnessed his meteoric rise to the level of astronomer, surpassing Hoge who remained at the observatory for years to come but was never promoted to the staff. Milt and Helen moved to a single-story home at 1034 North Hudson Avenue in a quiet garden section of Pasadena. With the influx of families moving to the area, Pasadena was becoming more civic minded. Plans were in place to build a civic center with a central library, city hall and large auditorium. The Rose Bowl had just been completed as a home to the Bruins of the recently founded University of California at Los Angeles. The newly renamed California Institute of Technology (Formerly Throop College) around the corner was just one of several good schools they could choose from when Billy was ready for college.

As they rang in the New Year, the family was celebrating the growing Witmer clan. David Witmer was back from the war and working as an architect in Los Angeles. Helen Witmer was pregnant with the couple's second child, while David's brother, William, and his wife, Esther, were showing off their newborn son, Edgar. Both of the Witmer brothers' first sons had been named after them. David and his young family were living in a Witmer designed home at 210 Witmer Street while his mother and sister lived next door to them at number 208. William and Laura Humason lived nearby at 323 Witmer Street.

At the observatory the focus of most of the departments and staff was on getting ready for the total eclipse in September. The instrument and optical shops were teeming with activity as specially designed interferometers and other equipment were tested and retested for use during the eclipse.

His work in the nebular and photometry departments kept Milt busy most of the year. On February 15, while on an observing run with the 100-inch telescope, he photographed the 22nd nova in Andromeda. The nova was located near the nucleus of the galaxy and would account for his only publication that year.

On July 1 George Ellery Hale announced his resignation as director of the Mount Wilson Observatory. The previous year spent abroad in an effort to recover from what his doctor called 'exhaustion' had not been helpful. Unable to sustain even a modest regimen without developing severe headaches and confusion Hale decided to step down. In the words of Adams, the director's decision "was received with deep regret by everyone associated in any way with its activities."

Among the men living and working at Mount Wilson only Milt had known it as long and as intimately as Hale. The two men had come there at very different moments in their lives, Milt as a wide-eyed 10-year-old boy and Hale as an established man of science, only 34 years of age, and already the founder of two of the most well-known observatories in the world. Nearly all of Milt's life had been spent growing up in the shadow of the observatory Hale had founded, and he was in awe of the exiting director. From his lofty vantage point, Hale had seen Milt grow from a boy into a man, marry the daughter of the observatory engineer and then somehow reinvent himself from janitor into an aspiring astronomer. He had to admit, Humason was acquitting himself quite well in his new position. On top of the work with Merrill on long-period variables, Milt accounted for nearly all of the photography of the Kapteyn Selected Areas and on spiral nebulae being compiled for comparison by Edwin Hubble, who had begun his classification of galactic and non-galactic nebulae shortly after his arrival in the fall of 1919 (Fig. 6.3).

Fig. 6.3 Milton Humason sits in front of a stereo comparator at the Mt. Wilson Observatory offices on Santa Barbara Street in Pasadena, California

Now, just as Milt was beginning his career, Hale was preparing to leave his beloved observatory. In a move that surprised no one, Hale chose Walter Adams to assume the position as Director of Mount Wilson Observatory, in charge of daily operations. Hale retained the position of honorary director, staying on to oversee "matters of general policy, the problems of research, and the development of new methods and new instruments." Free from the responsibility of the day to day running of the facility, Hale would have time to focus on his favorite subject, the Sun, for which purpose he had recently built a new solar lab on the grounds of his home in Pasadena.

On the same day that George Hale announced his retirement, Lewis Humason joined the Solar Department at Mount Wilson. For the first time since their childhood the brothers were together again. As a mechanic in a Field Artillery Unit during the war in Europe, Lewis had gained valuable experience tinkering with and fixing heavy and sometimes unruly machinery, always a welcome skill at a mountain observatory where quick fixes were often needed to keep instruments working while repairs were being made. Lewis was paired with Ferdinand Ellerman, the resident tinkerer and lead observer on the Snow and tower telescopes. As the family photographer growing up, Lewis assisted Ellerman in general photography at the observatory.

* * * * * * * * * * * * * * * * * **

On August 2, President Warren G. Harding suffered a massive heart attack in his hotel room in San Francisco and died. As Vice-President Calvin Coolidge was sworn in as Harding's successor, the country mourned the loss of a popular president who had restored peace and tranquility in the wake of World War I. Handsome and genial, Harding, who supported the Nineteenth Amendment, giving women the right to vote, had swept to victory in the election of 1920. His complacency and ignorance, though, in the prosecution of his office would become painfully apparent as corruption within his administration in the leasing of the government-held Teapot Dome oil reserve in Wyoming to well-paying business leaders was revealed.

The enormous effort by the staff in loading, moving and assembling the instruments for the solar eclipse on September 10 at Point Loma, Lakeside and on Mount Wilson, ended in disappointment. Extensive cloud cover over most of the Pacific Coast obscured the Sun's image during totality, yielding little usable information. The letdown in spirits was shared by members of other expeditions, who attended a joint meeting of the American Astronomical Society, the Astronomical Society of the Pacific and a detachment of the American Association for the Advancement of Science the following week.

Milt continued his work gathering spectra for Paul Merrill's research on long-period variables and the stellar parallax program of Walter Adams and Alfred Joy. In 1924 Merrill and Humason published "Discoveries and Observations of Stars of Class Be" together with Cora Burwell, who had assisted them in the computing department. It was the first comprehensive discussion of their work on these variable blue giant stars. The chief difference in the study of these stars from previous studies (as far back as 1866) lay in their ability to gain spectra of the line of emission rather than absorption in their atmospheres. The 30-page report culled data from their preceding six papers on the subject and revealed evidence of the stars' possible evolution into novae, the difference in the surface temperatures between these and other stars of the same class, the nature of hydrogen in their atmospheres, and their distribution in localized groups near the center plane of the Milky Way. Almost a hundred plates were made and studied for the paper, which was published in November, most of the plates made by Humason.

By the fall the family's eyes had turned to Lewis Humason as he prepared for his marriage to Beatrice Mayberry in February. Beatrice was the daughter of a well-known California family who had settled in the Alhambra area during the Gold Rush. In 1904, the family sold its El Molino Ranch to Henry E. Huntington through his real estate holding venture, who subsequently built a luxury hotel and golf course on the land near the foot of the San Gabriel range in Pasadena.

After graduating from Stanford University with a B.A. in physics in 1919, Beatrice had taught applied mathematics at her alma mater for a year before joining the computing department at the observatory. There she worked closely with George Hale and Seth Nicholson in solar research. The two had met shortly after Lewis joined the staff of Mount Wilson. Ginny and Billy Suermondt were in town for the event, having arrived in late September from Hong Kong aboard the luxury cruise liner, the S. S. *President Cleveland*. With their families gathered around them, Lewis and Beatrice were married at Episcopal Church of Our Savior in San Gabriel on February 7, 1924.

* * * * * * * * * * * * * * * * * **

As head of the stellar department, Walter Adams was eager for Milt to work with him and Alfred Joy on their program to study spectroscopic absolute magnitudes and parallaxes of M-type stars listed in the *Boss Catalogue*. Adams, who along with George Hale, George Ritchey (who had left the observatory in 1919 after years of disagreement and insubordination led Hale to ask for his resignation), Ferdinand Ellerman and Francis Pease was one of the original five members of the Mount Wilson Observatory, had distinguished credentials of his own. His recent discovery of the relationship between the relative intensity of certain lines in the spectra of stars and their absolute magnitude had won him the Gold Medal of the Royal Astronomical Society and the Henry Draper Medal for excellence in astronomical physics.

Adams reputation as a careful research scientist was as legendary as his frugality with all things concerning the operation of the observatory. Born to American missionaries in Syria during the Centennial year of the United States, Adams' family moved to Derry, New Hampshire, when Adams was 9, and he later attended Dartmouth College before moving on to the University of Chicago to earn a Master's degree in astronomy under the direction of George Ellery Hale. Adams later returned to Yerkes Observatory as Hale's assistant and followed Hale to Mount Wilson, where he worked with the solar telescopes until the 60-inch reflector was brought on line. As assistant director of the observatory, Adams consulted with Hale on nearly every aspect of the development of the observatory, acting as director during Hale's many absences for war work or the health issues that plagued the aging director. He was one of the most respected men on the mountain and kept the observatory on stable footing throughout his tenure as its director.

As working companions were concerned Alfred Joy was true to his name. Born in Greenville, Illinois, on September 23, 1882, Joy had graduated from Greenville University in 1903, earned his Master's degree in physics from Oberlin College a year later and served as director of the observatory at the Syrian Protestant College in Beirut from 1905 to 1914. A member of the Lick Observatory's expedition to Aswan, Egypt, to record the solar eclipse of 1905, Joy had spent a year at Princeton under a Thaw Fellowship and worked as a volunteer assistant astronomer at Oxford and Yerkes Observatory at the University of Chicago. He was on leave at Yerkes from Beirut, working as an instructor in 1914, when World War I broke out in Europe. With the outbreak of the war, Joy decided the smart thing would be not to return to Egypt and instead to seek employment elsewhere. He landed at Mount Wilson in 1915. It wasn't long before Joy partnered with Adams, immersing himself in the work on spectroscopic parallaxes. In 1920, he was named secretary of the observatory in charge of organizing the day-to-day operations, observing schedules of the staff and arbitrating the sometimes bitter disputes that arose between the astronomers at Mount Wilson.

A gregarious man with Old World charm and an insatiable appetite for science, Joy was the most productive astronomer in the history of the Mount Wilson Observatory. He excelled in observational, analytical and theoretical astrophysics, earning a Bruce Medal in 1950 for his contributions to astronomy. In spite of his tireless pursuits as a scientist, however, Joy seemed always to have time for those seeking his advice on an idea or a topic and was one of the most liked men on the mountain. A humble student with a desire to excel in astronomy could learn a lot working with such a patient and experienced man of science.

Due to their work in a non-research capacity on behalf of the observatory, Adams and Joy were in need of another teammate to handle the bulk of the observational work on the parallax program for M-class stars begun in 1923. Adams would have to devote much more of his time to the details of the everyday operations of the facility while he made the transition to full-time director. The Carnegie Board must be assured that Adams was up to the task of running the observatory, even though he came to the position with Hale's full endorsement, and most of the board members knew Adams and had worked with him in recent years while Hale was away. Seeing the promise Milt had shown working with Paul Merrill, Adams decided to asked Humason to join in the work. Milt leaped at the chance to work with Adams, who had been so instrumental in his advancement. In 1923, Milt set about the task of gathering spectral data on the selected stars for the program. For the next two years he compiled the physical data, recording information on their absolute magnitudes that would later be used to determine each star's spectral type.

It was during this period that Milt started to hone his special talent for stellar photography as he worked alongside Alfred Joy. Most of the stars Milt was fishing for were very faint and not visible to the naked eye, making them difficult targets. Photographing these objects took not only competence in the science of tracking them but a comprehensive knowledge of the idiosyncrasies of the instruments he was using. Like all machines, the large reflectors at Mount Wilson had their little glitches. To compensate for these, the observer had to know what to expect in order to stay ahead of the problem. The better that person was at managing these tiny imperfections the smoother the telescope would operate and the sharper the exposure.

The 100-inch telescope had its massive weight and movement balanced and supported on mercury floats. Mercury was very expensive, and the observatory, operated as it was on a stiff budget by the frugal Walter Adams, often ran low on its supply of the liquid lubricant. The level of mercury in the tanks had to be kept high to keep the telescope moving smoothly. At lower levels the mercury in the tanks would heat up, causing a foam to develop that would fill the empty space in the float chamber. An abundance of the foam would cause slight vibrations in the movement of the telescope that would translate into blurred images or spectra. The fainter the object, the longer exposure time was needed and the greater chance that the astronomer would have to deal with the problem in the floats. This was a bad thing.

As a result of this effect, observers and night assistants alike could be found scraping up small pools of mercury from the observing platform to try to preserve what they could for use in the floats. As his knowledge of the intricacies involved in the telescope's operation increased Milt began to predict the problems that would arise and to counter them with a higher and higher degree of precision. As a result he was able to take long exposures of increasingly distant and faint objects with greater success.

In 1925, Adams, Joy and Humason had gathered what data they needed for the catalog and started work on their system for assembling the stars into spectral types from M0 to M7. The spectral type of the stars was distinguished by increases in the band of titanium oxide in their spectra. The coded table was designed to give an objective glimpse at each star's temperature and density. After a year's work carefully choosing a class type for each star the team published its first work on the subject in the *Astrophysical Journal*.

During the year George Hale donated his new solar laboratory in Pasadena to the observatory as part of the Carnegie Institution of Washington. That year Edwin Hubble, using distance scales derived from Cepheid variable stars he had found while studying earlier plates, first reported his finding that Andromeda was actually a galactic system of its own. The new distance to Andromeda given by Hubble's measures was 900,000 light-years and a diameter of 50,000 light-years. It was one of the most significant discoveries in science in years and would lead to later collaboration between Hubble and Humason, discussed in the next chapter.

As the year ended, Lewis and Beatrice celebrated Christmas that year with the birth of their only child, Harry Wing Humason, born on Christmas Eve, 1925. The boy got his unusual name from his great grandfather, George Wing, who had created a system of function symbols to help teach essential sentence structure to deaf and mute students in 1884. Wing's work influenced the likes of Alexander Graham Bell, who was also active in helping the hearing impaired. His system of symbols was later referred to as Wing symbols and is used in schools for the hearing impaired to this day. With the birth of the boy, Beatrice was now at home while Lewis worked at the observatory.

The 1926 joint report on absolute magnitudes and parallaxes laid out the course of study of the abundant red dwarfs and giants that made up three quarters of all main sequence stars. The team illustrated the close agreement between their spectroscopic magnitude method and the trigonometric method that had long been used to determine stellar magnitudes. The majority of the stars were red dwarfs with a smaller portion of them being red giants. Twenty-eight of the stars were referred to by Adams, Joy and Humason as "super-giants," a term used regularly today to describe stars such as Betelgeuse in Orion.

In January, Albert Michelson conducted tests to determine the velocity of light with members of the observatory optical and instrument shops. Using specially designed equipment and mirrors, the team set up two stations between Mount Wilson and Mount San Jacinto at a distance of 90 miles. The exact distance between the apparatus at the two stations was determined with high precision, the light passing between the stations at over 500 revolutions per second. Although storms and haze interrupted their work Michelson and team were able to derive a mean velocity of nearly 300,000 km/s after several successful tests.

As the number of different research programs increased Milt was contributing to nearly every phase of the nebular department. By now his ability to photograph readable spectra of exceedingly faint objects was becoming obvious. In April Milt photographed the spectra of Wolf 359, by far the faintest star known at the time, and derived a radial velocity for the star of −90 km/s.

The success of his debut collaboration with Adams and Joy on class M stars only increased Milt's appetite for learning all he could about the science of astronomy. An eager student he had the benefit of tutelage from fellow collaborators such as Seares, Adams, Joy, Merrill and Nicholson, all of whom benefitted from his special skills as a photographer. As a technician at the eyepiece of the Mount Wilson telescopes he was gaining respect and admiration among his peers. The spectros-copists of Mount Wilson had taken their measure of Milton Humason, and their unanimous support for his abilities set the stage for his extraordinary rise in the field of astronomy.

Chapter 7
The Battle Over the Structure of the Universe

Abstract Blessed with great natural skill at the telescope Milt finds that his talents are his enemy when he takes on an incredibly long observing run to photograph the spectra of extremely far off galaxies for Edwin Hubble. Unhappy with the notion of continuing with the work, Milt quits the project and it takes an invitation by the honorary director of the observatory, George Ellery Hale, and a promise of innovations in camera technology to change his mind. The new technology proves equal to the task, reducing exposure times for deep space galaxies and helping Hubble and Humason to push their velocity-distance relationship farther and farther, until Einstein himself is ready to concede that the universe must be expanding. The world's best-known scientist visits the observatory and meets with Hubble and Humason to discuss their findings.

Less than a hundred years ago most of the world went about its business believing the whole of the universe existed within the Milky Way Galaxy. The sky, resplendent in starry richness and cradled in a sea of darkness, contained all of creation, marvelous in scope and streaked through with a broad strip of gauzy light that outlined the shape of the galaxy.

Within the science world, however, a war was developing over the formally accepted size and nature of the universe, one that had been shaped by centuries of discovery and debate, intellectual zeal, public humiliation, punishment and sacrifice. Men such as Hipparchus, Ptolemy, Copernicus, Kepler, Galileo and Newton had spent their lives in the pursuit of understanding the structure of the universe. Many others, like Tycho Brahe, Charles Messier, William Herschel and William Huggins, had begun to unravel the mysteries of the stars, identifying and cataloging their place in the sky. One of the key issues of debate on the structure of everything formed around the question of whether the stars were part of the Milky Way or, in some cases, massive groups of stars in their own right, which the German philosopher, Immanuel Kant, described as "island universes" in 1755.

The word "galaxy" comes from the Greek word *galaxias,* meaning "milky," so to some extent the name of our home galaxy is redundant. The Milky Way Milkiness would have been a weak title for the cradle of all creation, however. In any case, the word given our starry home evokes an image of smoothness that

© Springer Science+Business Media New York 2016
R.L. Voller, *The Muleskinner and the Stars,*
Springer Biographies, DOI 10.1007/978-1-4939-2880-4_7

would have described well the distant smudges seen by early stargazers. In 1845 a man named William Parsons, the 3rd Earl of Rosse, built what might be the most aptly nicknamed telescope of all time. Known as the Leviathan of Parsontown, the behemoth, with its 72-inch mirror, was the largest telescope in the world until the Hooker Telescope at Mount Wilson went into use in 1919. (Although smaller than Rosse's telescope, the 60-inch reflector was far more advanced technologically and far more useful for astronomical observations.) Using this giant reflector, Parsons was able to resolve M51 (the Whirlpool Galaxy) into its spiral shape. Parsons claimed that this and other nebulae could be resolved into stars.

In 1915 Albert Einstein published his General Theory of Relativity, furthering his conception of the nature of light and gravity. The new theory, which followed his earlier paper on special relativity, elaborated on the theme of space-time and considered how the path of light might be altered by objects having extremely high mass. Two years later Einstein decided to use his new theory to describe the size and shape of the universe, which he assumed would fit neatly into the common notion that it was static and unchanging. Instead he watched in dismay as the universe collapsed under its own weight with every attempt at a workable solution. Not able to stomach the idea of a shrinking universe, Einstein came up with a carefully chosen figure for his general theory to keep the universe in a Steady State. He called this new figure the cosmological constant.

For the moment anyway Einstein's constant had saved the universe from ultimate ruin, and stargazers the world over could go on happily staring at the twinkling light above secure in the notion that the heavens were eternal, infinite and unchanging.

In 1918, Harlow Shapley published a series of papers that would alter some of these perceptions among many in the scientific community. The fourteen-part series, the culmination of Shapley's study of stars in globular clusters from his work with the 60-inch telescope at Mount Wilson, was an observational breakthrough that set the stage for a debate on universal structure two years later. Shapley's conclusion, based on four years of research, was that the Milky Way contained all the known stars and matter in the universe, that it took the form of a disc-like spiral, like a pinwheel, and that the Sun occupied a place in one of the spiral arms of the galaxy, far from its center. He went on to conclude that, based on his calculations of the distances to Cepheid variables in some of the clusters, the galaxy had a diameter of roughly 300,000 light years.

Shapley had based much of his work on a distance scale that resulted from the efforts of a Harvard "computer" named Henrietta Swan Leavitt. Shortly after her arrival at Harvard in 1893, observatory director Edward Pickering had given her the task of studying variable stars whose luminosity varies over time. Leavitt, the daughter of a Protestant Church minister, took to the work with a puritanical zeal. In 1912 Leavitt published her findings, which concluded that some of the stars, so-called Cepheid variables, in the nearby Magellanic Clouds were brighter and had longer periods than others, and that their period and luminosity were predictably similar. Leavitt's period-luminosity relationship for Cepheid variable stars was the lynchpin in the freight train through the cosmos. Armed with this information

Danish astronomer Ejnar Hertzsprung calculated the distance to several Cepheids in the galaxy in 1914, setting forth a system for determining the distance to these stars anywhere in the galaxy based on Leavitt's period-luminosity relationship and each star's apparent magnitude.

Harlow Shapley began using Hertzsprung's formula in 1915 to reposition the Sun and determine the size and shape of the galaxy. For the first time, astronomers had a means of measuring approximate distances for deep space objects. On assuming the directorship at Harvard Observatory in 1921, Shapley immediately promoted Henrietta Leavitt to head of the photometry department. The appointment was short-lived, however, as she died of cancer on December 12 of that year. Her beloved Cepheids would continue to help shape the course of astronomy for years to come.

Shapley's universe was spacious by almost any standard. After all, it would take a man driving a Bugatti, the fastest car in the world at the time, at its top speed of roughly 60 miles/h, about a 100 billion lifetimes to go from one end to the next. That's one heck of a road trip! The standards of the island-universe crowd were different, though. To this growing group of scientists Shapley's assertion that all matter existed within the Milky Way was untenable and would soon be overturned in favor of a universe full of galaxies like our own. Shapley was defiant in the face of this pressure. In letters to Hale, Shapley unveiled his vision of the universe in soaring flights of empirical conjecture, deduced from his work on globular clusters and other stellar data as well as van Maanen's photometric measurements of stellar parallaxes. A native of Missouri, born in 1885, Shapley had a driving ambition fueled by an enormous ego that sometimes clouded his judgment. In spite of the albeit highly educated opinions of his young prodigy, Hale remained skeptical that there wasn't more to be learned about the nature of the heavens. Writing in reply to one of Shapley's more romanticized letters on the nature of the universe, Hale wrote: "I confess that I still entertain many doubts about the nature of the spiral nebulae and their relationship to our own system. I also feel a bit skeptical regarding your hypothesis to account for the novae in spirals. We must evidently give more and more attention to all phenomena relating to the spirals and accord them a large place in our revised observational program."

Two years later, in a seeming slight to Shapley, Hale publicly revealed his ambivalence on the subject in an article in Scribner's saying: "The spiral nebulae, more than a million in number, are vast whirling masses in process of development, but we are not yet certain whether they should be regarded as 'island universes' or as subordinate to the stellar systems which include our group of minute sun and planets, the great star clouds of the Milky Way, and the distant globular star clusters."

Around this time in 1920 Hale came up with an idea for a debate on the subject to be held at the National Academy of Sciences in Washington, D.C., as part of the William Ellery Hale Lecture Series, named for his late father. The debate on "The Distance Scale of the Universe," would be argued on either side by Shapley and Hebert Curtis of the Lick Observatory. After the debate each man would assume directorship of an observatory of their own, Shapley at Harvard and Curtis at the Allegheny Observatory of the University of Pittsburgh.

The debate was held on April 26, 1920, with each side making its case before a jury of its peers. It ended in a draw, with the proponents from both sides walking away unswayed by the passionate arguments of the other. Even more sobering was the fact that the customary wine served at the end of each lecture was not to be poured, as the great experiment, Prohibition, had gone into effect in January. Had Shapley been as good behind the lectern as he was at the written word he might have fared better in the debate with the more polished orator, Curtis. In a crowd the youthful-looking Shapley had a tendency to become a bit sheepish under scrutiny. Knowing this about himself, Shapley had even tried to dissuade Hale from organizing the debate in public, preferring a 'discussion' in private between he and Curtis with Hale and Lick Observatory director, W.W. Campbell presiding.

The buzz over the debate receded into history and the third decade of the twentieth century began without any closure on the nature of the universe. Although most of the scientific community agreed Shapley was right in his conclusion on the location of the Solar System, itself a major breakthrough, no consensus had yet been reached on the bigger questions of island versus single universe. The answer would come from another Missouri native who joined the Mount Wilson Observatory just about the same time the Great Debate was held.

* * * * * * * * * * * * * * * * * **

Edwin Hubble came to Mount Wilson in the fall of 1919, in every way the embodiment of George Hale's ideal astronomer. He had been an undergrad at the University of Chicago, a Rhodes scholar who studied at Oxford from 1910 to 1912 before returning to California to earn his Ph.D. at Yerkes Observatory. Drafted into the army during World War I, Hubble finished his thesis just as he was shipping off to Europe.

The young and self-consciously virile Hubble was a superstar during his college years at the Yerkes Observatory, where he used the Ritchey 24-inch reflecting telescope to hint at his life's work on a new system for categorizing nebulae. Born November 20, 1889, in Marshfield, Missouri, the pompous purveyor of the night skies had an affected English accent he picked up during his time at Oxford, often wore jodhpurs and a smoking jacket and insisted on being referred to by his rank of major even though he had seen no action during the war. Apparently shy around colleagues that he spent little time with, Hubble would invariably try to control social events that might otherwise be awkward to him. When dining with colleagues on the mountain, the standard social practice during observing runs, Hubble got into the habit of reading up on some obscure subject he would find in the *Encyclopedia Britannica* and hold forth on the subject during dinner.

Hubble strode onto the grounds of the Mount Wilson Observatory just as his predecessor, Harlow Shapley, was ending his historic career to lead the Harvard Observatory into the twentieth century. His arrival also came as the 100-inch Hooker Telescope was coming on line. Armed with two of the world's largest telescopes Hubble soon began his campaign on nebular research.

One of the first staff members to work with Hubble was Milton Humason, a night assistant who was taking his first tentative steps in astronomy. Shortly after Hubble's death in 1953, Humason recalled his first memory of Hubble from 1919, when Hubble was at the eyepiece of the 60-inch telescope:

> He was…standing while he did his guiding. His tall, vigorous figure, pipe in mouth, was clearly outlined against the sky. A brisk wind whipped his military trench coat around his body and occasionally blew sparks from his pipe into the darkness of the dome. "Seeing" that night was rated extremely poor on our Mount Wilson scale, but when Hubble came back from developing his plate in the dark room he was jubilant. "If this is a sample of poor seeing conditions," he said, "I shall always be able to get usable photographs with the Mount Wilson instruments." He was sure of himself – of what he wanted to do, and of how to do it.

One can forgive the romantic license taken by a friend writing in memory of a fallen cohort, but Humason's recollection likely hides what must have been at best some ambivalence upon first getting to know Hubble. At first glance it might be said the two were diametrically opposed. If Hubble was combative and contentious, Humason was forgiving and patient. Hubble was driven by ambition, and Humason was amiable and humble. Hubble was quiet, intimidating and introverted whereas Humason was the life of the party and got along with everyone. Hubble was ostentatious and Humason was down to Earth. Hubble smoked tobacco and Humason chewed it. Hubble preferred fine wine, scotch or sherry, and Humason drank homemade moonshine. Nearly every facet of their personalities was different from the other.

Hubble's affected British air and demeanor, and his penchant for uttering 'what?' at the end of his sentences must have made Humason grimace in disapproval in the early going. As he later learned, though, Hubble had simply adopted England as his own, and by most accounts the feeling was reciprocated. During World War I Hubble was a fierce advocate for England and was known to strongly rebuke anyone not interested in aiding his adopted countrymen in their fight against tyranny and dictatorship. Whatever he felt of Hubble when he first cast eyes on him, Humason grew to admire him not only for his brilliance as an astronomer but for his character as well.

* * * * * * * * * * * * * * * * * **

In 1922 Hubble had published two papers on the structure and makeup of both galactic and non-galactic nebulae. In these papers Hubble laid the groundwork for a new classification system for the nebulae, breaking them down into several broad groups, including spiral, elliptical, elongated, globular and diffuse. His system set out to capture all the nebulae into large groups that could subsequently be categorized more specifically. He proved that their brightness was due to the existence of a bright star of B1 or an earlier type that Humason and Merrill had been reporting on in their series on Class-B stars. His conclusions also proved the existence of dust as a reflective property of nebulae, meaning they were made of space stuff of some kind or another and not just wispy clouds of gas.

Another area of interest to Hubble was the study of novae. Milt had recently discovered one of these exploding stars in Andromeda on June 23 of that year. Hubble now undertook to study these in greater detail to see how he could integrate his findings on them into his understanding of the nebulae. During the course of this study Hubble made a groundbreaking miscalculation, one that would reverse Shapley's single universe theory and end the Great Debate. In a plate made at the 100-inch telescope on October 5, 1923, Hubble incorrectly identified an object in M31 (Andromeda) as being yet another nova. He later realized that what he had found was not a nova at all but a Cepheid variable.

Hubble did some digging and found Cepheid variables in M33 and NGC 6822 as well. Using Shapley's method of calculating distances from these unique stars he quickly learned that the universe was getting larger—a lot larger! At a distance of almost 3 million light years, Andromeda was 10 times more distant than Shapley's universe could encompass. Therefore, Hubble concluded, M31 and other nebulae must be distant universes made up millions of stars like those in the Milky Way. The island universe theory was confirmed. After working to corroborate his evidence for the next several months, Hubble first published his findings in *The New York Times* on November 23, 1924, and again in a formal paper at the meeting of the American Astronomical Society on January 1, 1925. The discovery made national news and Hubble a sensation overnight, bringing further renown to the observatory.

For ambitious Hubble, the breakthrough allowed him to step out of the enormous shadow Shapley had left behind and into the spotlight he so coveted. Neither man was a fan of the other anyway, but Hubble had been uneasy about Shapley's high position within the science community from the moment he had arrived at Mount Wilson. Eager to deflate Shapley's balloon, Hubble sent him a letter giving him the news of the Cepheid in Andromeda first hand: "You will be interested to hear that I have found a Cepheid variable in the Andromeda Nebula (M31). I have followed the nebula this season as closely as the weather permitted and in the last five months have netted nine novae and two variables."

The letter, which Hubble addressed to his equally egotistical rival in February of 1924, was written with the intention of pouring salt into Shapley's wound. Upon reading the letter Shapley was heard to utter, "Here is the letter that has destroyed my universe."

Such was the nature of the rivalry between Hubble and Shapley. It would remain contentious for the remainder of their professional lives, Hubble continually seizing the initiative and Shapley vehemently denying the evidence no matter how universal opinion was against him. Like most things in life, fate played a role in the fortunes of both men. Had Shapley had the 100-inch telescope at his disposal during his time on Mount Wilson, he would have had the advantage of greater clarity and photographic power that Hubble enjoyed after 1920. In the end Shapley would fare well as the director of the Harvard College Observatory, serving that institution for over 30 years and placing it on firm footing through the middle of the twentieth century. Among the Ph.D. students Shapley mentored during his time at Harvard were Cecilia Payne-Gaposchkin, who was the first to suggest that hydrogen was the predominant element in the universe, and George Lemaitre, the progenitor of the Big Bang Theory that would bring Hubble and Humason together in an historic partnership toward the end of the decade.

If Milton Humason held any ill will toward Edwin Hubble in the early years of their association he wouldn't have shown it. Much too conscious of his place as a scientific outsider among the staff Humason was more apt to throw himself into the work going on around the observatory and leave the bickering and infighting to others. His backwoods charm and social guile were the perfect tools for building long-lasting working relationships with members of all departments at Mount Wilson. As his admiration and respect for Hubble grew through their close association, Humason would be torn between his loyalty to Shapley, whom he always considered to be his greatest champion during his early development, and Hubble, with whom he was pioneering in the realm of sidereal evolution. As time wore on, in fact, he had more reason to doubt his mentor's competence and foresight. Shortly after Hubble's discovery of the Cepheid variables in Andromeda, for instance, Humason found out that a plate Shapley had made on the 60-inch telescope in 1919 was used by Hubble to corroborate his evidence. During his mentorship with Shapley in 1919, Humason was given some plates of novae Shapley had taken to study for his work on the single universe theory. Humason used a blink comparator, an instrument that used flashing light swiftly moving between two plates of the same region of space, to study their rotations and other attributes. While studying some plates of M31 he discovered what appeared to be variations in the brightness of one of the novae. He carefully circled the object with a wax pen and brought the plates to Shapley to get his opinion. Could it be that they had resolved images of a Cepheid variable in Andromeda, he wondered outloud. Shapley sneered at the notion that the object was anything more than nova as he pulled a handkerchief from his pocket and wiped the plates clean, erasing the marks Humason had made. A novice at the time, Humason was in no position to argue with his older and extremely accomplished mentor. Had he stood his ground and demanded Shapley take a closer look at the objects on the resolved plates, Humason may have helped to change the course of recorded history. This event and the results of his research with Hubble would eventually lead Humason to distance himself from Shapley's arguments, although he remained a friend to his early mentor throughout his life.

* * * * * * * * * * * * * * * * * **

As 1925 began news that the size of the universe had increased exponentially was sweeping the world. Just how big was still under some scrutiny, and some, like Shapley, were still denying Hubble's claim, but most of the world was mesmerized by the thought that the universe was made up of what might be millions of galaxies similar to the Milky Way. And while the news was spreading that the universe was seemingly infinite a new ripple was starting to be felt within the science world.

Working from his home in isolation starting in 1917, a Russian-born mathematician named Alexander Friedmann began work on his own theoretical version of the universe based on Einstein's general relativity equations. With Russia caught in the throes of a world war in Europe and a revolution at home, Friedmann wasn't aware of Einstein's cosmological constant, and so his early model ran the same line of contraction that caused Einstein to invent it in the first place. As the dust from international and civil conflict settled Friedmann learned of Einstein's 'fudge factor' and began to apply it to his model. In 1922 he published a paper in the *Zeitschrift für Physik* that described different solutions to the fate of the universe based on changing values for the cosmological constant, from an ultimately collapsing universe to one that remained in a steady state or one that continuously expanded.

After first decrying Friedmann publicly, Einstein acquiesced and admitted in writing that the young physicist's revisions on general relativity were "both accurate and clarifying." However, Einstein continued to endorse his version of the static universe and considered Friedmann's additional solutions ultimately meaningless. Unfortunately, Friedmann never got to see the outcome of his work. He died of typhoid fever in 1925 at 37 years of age. The timing of his death and his Copernican-like activism probably kept Friedmann's theoretical brilliance from reaching the public sphere for many years. His principals would soon be revived in the hands of a man whose insight, stature, and charisma would begin a revolution in science.

* * * * * * * * * * * * * * * * **

Georges Lemaitre was ordained as a priest in Maline, Belgium, in 1923. A graduate of the University of Louvain prior to his entering the seminary at Maline, Lemaitre next spent a year studying physics with Arthur Eddington at Cambridge and then a year studying astronomy at Harvard College under the tutelage of Harlow Shapley. In 1925, he returned to the University of Louvain to accept an academic post while he finished his Ph.D. in physics.

Unlike other notable contemporaries, Lemaitre was a hero of the Great War, having earned the Croix de Guerre for bravery under German poison gas attacks. When asked how he balanced science and religion Lemaitre was known to reply, "There were two ways of arriving at the truth. I decided to follow them both."

In his own researches, done independent of Friedmann's earlier work and apparently without knowledge of it, Lemaitre combined Einstein's equations with the most up to date physical evidence of the day to build a theoretical model of the creation and expansion of the universe from the explosion of a primeval atom and evolving until present day and beyond. He published his theory in a paper entitled "Hypothése de l'atome primitive" in 1927. In response to Lemaitre's assertions, Einstein rebuffed the Belgian priest and physicist as stiffly as he had Friedmann, saying to him at the 1927 Solvoy conference in Brussels, "Your calculations are correct, but your physics is abominable."

In spite of Einstein's attempts to derail Lemaitre's theory, word of the expansion principal made its way around the scientific community until it reached the ears of Edwin Hubble, who had been scheduled to chair the meeting of the International Astronomical Union at Leiden in 1928. At the conference Hubble met with a group of astronomers who mentioned the expansion predicted by Lemaitre's equations. If Lemaitre was correct and the universe was expanding then the nebulae should be receding from Earth. The further away the nebula the fainter it should be and the greater its redshift, or velocity through space.

By the time he got back to Pasadena in the early summer, Hubble was eager to find out if Lemaitre's predictions could be confirmed observationally. The trouble was that the objects he was going to have to target for his research were going to be difficult if not impossible to get in terms of spectra and other usable data. If he was to have any success Hubble would need to find the best stellar photographer on the mountain, a guy with patience, skill and, preferably, someone who would not try to steal his thunder. Hubble went to Walter Adams to seek his opinion on who among the staff might fit his criteria. Adams had just the guy.

* * * * * * * * * * * * * * * * * * **

By 1928, Milton Humason was quietly becoming a star on Mount Wilson. On the heels of his major publication with Adams and Joy on absolute magnitude and parallactic classifications of M-type stars Humason was becoming well known among the staff for his ability to get clear exposures of objects nobody else could get. Where others failed Milt was getting good sharp images and spectra of deep space phenomena using the large reflector and winning praise for his work in the process. The all-encompassing work on stellar parallaxes with Adams and Joy had taken up the majority of Humason's time the past two years, ending in the publication of their results in 1926 (an addendum was made to the paper in 1927).

In 1927, Humason published four papers, all on separate topics and two with different collaborators. The first, entitled "Note On Very Cool Stars," was written with Paul Merrill. Seizing upon the evidence compiled in both the B-type and M-type star programs, Merrill and Humason set out to paint a picture of the current state of understanding of stellar evolution. Having done most of the heavy lifting on both of these programs, Humason was in the best position to weigh in on the observational aspects of the topic. The key questions, Merrill and Humason pointed out, were whether still cooler stars existed, and if so, how these stars might be identified and whether stars of the proper surface temperature could be found to account for a gradual falling off of temperature and density to fill in the timeline of the lives of various types of stars (Fig. 7.1).

This was the kind of stuff that might make those who weren't interested in the minutia of stargazing want to go to the beach or, in Humason's case, fishing. It was the kind of research that turned a guy like Merrill on, however, and Milt, being generally interested in the idea of stellar evolution and in helping out his friend and colleague, Merrill, went happily along for the ride. After all, he'd spent many a long night on the mountain tediously inching his way across the heavens in the pursuit of the evidence; he might as well stick around and see what exactly it was that he'd caught.

By this time, Hubble's rise in popularity was beginning to rub the spectroscopists at Mount Wilson the wrong way. Friends at the onset, Merrill and Hubble were steadily growing apart, their mutual conservatism (they were the two most conservative members of the staff) wasn't enough to make up for the differences in their

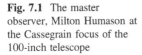

Fig. 7.1 The master observer, Milton Humason at the Cassegrain focus of the 100-inch telescope

approach to science. Merrill had become critical of Hubble's less than detailed approach to research. Hubble's cavalier ways and deductive genius had merely grown the universe from the unfathomably large to the absurdly unfathomably large. For his part, Merrill still sided with Harlow Shapley, who denied Hubble's conclusions on the distance scale to the nebulae based on the work of his friend, Adrian van Maanen, an expert in the field of stellar parallax at Mount Wilson, whose measurements of the galaxies in Hubble's conclusive paper were in line with Shapley's belief that they couldn't possibly be as far off in space as Hubble had claimed. In arguing with Merrill on the merits of Shapley's and van Maanen's results and their stubbornness in the face of growing scientific and public acceptance of the new scale of the universe, Hubble told Merrill that any friend of Shapley and van Maanen was no friend of his. Hubble's attitude suited Merrill just fine.

The rift between Hubble and van Maanen boiled over one evening at dinner on Mount Wilson, where it had become customary for the operator of the 100-inch telescope to sit at the head of the table at meal time. The 60-inch operator would sit to one side and the solar tower operator to the other side. Van Maanen, there on a run with the large reflector, dressed and readied himself for dinner in his room at the dormitory known as the Monastery. Hubble was on the mountain for an observing run on the 60-inch telescope and was entertaining friends whom he had invited to sit with him at table.

Van Maanen walked from his room and made his way through the main corridor to the dining room, chatting idly with a friend as he approached the table. He could not be prepared for what he was about to find. Hubble was seated at the head of the table, his friends sitting beside him. Stupefied, but unwilling to provoke the formidable Hubble, van Maanen found his name plate a few places down and humbly took his seat.

Humason, characteristically, was working with all of these men on various projects. His next paper was written with Adrian van Maanen and J.A. Brown and was entitled "A Star of Extremely Low Luminosity." The star in question had been on the parallax program of Adams, Joy and Humason, and van Maanen's investigation was to determine the star's proximity and intrinsic brightness. Brown had made five plates of the star and determined its photo-visual absolute magnitude to be 16.5, making it the faintest star ever recorded. Knowing the chances were slim of getting a good result, van Maanen asked Humason to try anyway to get a good spectrogram of the star for classification and a measure of its velocity through space.

On April 23, 1927, Milt went up to the top of Mount Wilson. It had been a snowy April, 17 inches of snow had fallen on Los Angeles alone and the mountain road was even more treacherous than usual. Later that night Milt stood in the darkness on the cold observing platform as the enormous dome rotated until the slit was in position for his observing run. Seated at a wooden chair Milt peered into the great void, marking a guide star he hoped would help him track the star in question and get him a successful spectrum. The spring night air was crisp, but the seeing was good, and Milt thought he had a good chance at getting the result he wanted. Working the controls Milt carefully moved the giant reflector and the dome across the sky chasing the object. After 15 min he closed the shutter and removed the plate, inserting another before checking his position and starting again. Over and over that evening he repeated this process until he had several exposures of the star's spectrum to choose from. He only needed one to work. The next day Milt processed his spectrograms and chose the one he liked best for van Maanen's study. He calculated it as having a closing velocity of 90 km/s. Milt handed the slides and his measurements to his partner, who was impressed and pleased at having chosen Humason for the job.

It was around this time that Humason made his first foray into nebular work. The fury unleashed in January of 1925 after Hubble's announcement of the Cepheid variables in Andromeda and his distance scale to galaxies of 1926 was palpable on the grounds of the observatory for many months. Conversations at the poker table and over billiards in the Monastery basement all revolved around Hubble's new universe. The spectroscopists, many of whom sided with Harlow Shapley, were eager to find fault with Hubble's assertion that the nebulae existed outside the Milky Way. The spectroscopic absolute magnitudes of stars given in Walter Adams program showed half the magnitude of Hubble's measurements. Van Maanen's internal motions of local galaxies were also revealing shorter distances than Hubble's new distance scale suggested.

George Hale, still actively guiding the observatory's direction from his home in Pasadena, decided more proof must be supplied to give Hubble's theory real evidence to either corroborate or discount his claim. To achieve that end Hubble needed someone to help him with his program on nebular classification, someone Hubble could get along with and whose skill could be trusted to gather the evidence required. To Hale's surprise the man Adams chose for the job was the little known former muleskinner, Milton Humason, the devoted friend and former protégé of Hubble's arch rival, Harlow Shapley. What was Adams thinking? Adams had a hunch that although Humason's allegiance was squarely in the court of Shapley, he would ultimately give credence to any solution that seemed real and support solid evidence. Humason, he knew, gave everyone a chance to prove himself.

* * * * * * * * * * * * * * * * * **

Late in 1926 Humason began taking direct photographs and spectra of NGC 6822 (Barnard's Galaxy), an oddly shaped barred galaxy that was part of Hubble's first single universe-crushing paper. Over the next several months Humason took two plates of the galaxy before turning his sights on M101, the Pinwheel Galaxy. It is at this point that he pulled off one of the most daring stunts in Mount Wilson history.

The Pinwheel Galaxy is located in the tail of the constellation Ursa Major and crosses its meridian point at a declination of +54° North. The 100-inch telescope was mounted on a closed yoke equatorial mount with bearing houses at the north and south end. The massive north mounting bearing sat on a high pier while the south bearing was mounted lower so that the yoke was mounted on an angle according to the design by Francis Pease. Photographs of objects such as M101 required the use of the Newtonian focus cage, whose secondary mirror was mounted near the front of the tube. To photograph M101 at meridian the cage must be mounted with the eyepiece facing north, toward the main bearing, which was filled with mercury, used for lubrication at the time. Crossing the meridian north meant that the observing platform (with the observer perched upon it) had to pass over the north bearing. This was not a problem until you reached about +50° North, at which point either the observer got mushed like a pancake between the enormous telescope and the viewing platform, was swept from the platform onto the concrete floor below or, worse yet (to the astronomers anyway), the viewing platform slammed into the north bearing, perhaps rupturing a seal and triggering a major mercury event. The resulting ignominy and opprobrium was enough to make most men (including Hubble) shy away from any attempt at reaching M101 at meridian passage. Not Milton Humason.

Humason made his first attempt in May. Working through the approximate position of the platform and telescope at the high point above the north pier he confirmed that they would fit without hitting each other or the pier. The question was, could he fit, too? Shortly after nightfall he began his observing run. Laying on his back, his hands stretched above his head to reach the telescope controls, Milt gingerly worked the clock and dome rotation as he tracked the galaxy in the cold pitch darkness for hours, easing the platform over the north bearing housing while contorting himself to avoid pinching his fingers (or himself) between telescope and platform, feet dangling 30 feet above the concrete floor of the dome. The night assistant, Tom Nelson, stood by, shocked and amazed at the composure and skill shown by his old friend as Milt worked himself, the giant reflector and dome across the night sky.

When he had finished, Milt hopped down from the observing platform, plate in hand, and looked at his awestruck friend and night assistant. "Not bad," he said, "I'll come back in June and try it again."

Humason published his results in the Astronomical Society of the Pacific's journal in September of 1927. In the paper he described the faint "knot of nebulosity" characteristic of the galaxies and reported on the detailed results of the 4- to 6-h exposures. He compared his spectra to those taken by Hubble in 1923, saying the definition on Hubble's plates, "was not sharp," and concluding: "The radial velocities of both NGC 6822 and Messier 101 are unusually low for non-galactic objects. This is consistent with the marked tendency already observed for the smaller velocities to be associated with the larger (and hence probably closer) nebulae and those which are highly resolved."

At the completion of this latest paper, Humason put down the nebular work and went back to a project more to his liking, namely, the study of novae. These distant remnants of exploded stars were of great interest to Milt, and, as he had discovered one or two of his own, he was excited by the possibility of finding more and uncovering their secrets. In November of 1927, Humason published a five-page report on Nova Aquilae in the journal of the Astronomical Society of the Pacific detailing the results of several months of study of the nova's declining luminosity and other attributes.

Word of Milt's heroics in photographing M101 spread like wildfire. At his office in Pasadena, Edwin Hubble soon got wind of Humason's feat and decided this must be the man to help him take on his assault of the extra-galactic nebulae (Hubble's term for galaxies that would stand at Mount Wilson and Palomar until after his death, when the term "galaxy" was adopted). Guided, as he was, by his ego and ambition Hubble would have viewed the prospect of working with the most skilled technicians as a rite of passage, a feather tucked neatly into his intergalactic cap.

On that fateful day in the summer of 1928, Milton Humason was working in his office when he got a surprise request to come to Hubble's office. Unsure of what to make of the invitation Humason slowly padded down the hall to Hubble's door and knocked. After a moment the door swung open with Hubble on the other side of it, smoking a pipe and eyeing Humason with a slight smile. Wearing his best poker face, Humason entered and took a seat behind Hubble's desk. As Milt sat in Hubble's office subsumed by the Hubble mystique, the juxtaposition of characters was palpable—Humason seated near Hubble's desk, chewing tobacco, a flask of homemade moonshine in his jacket pocket, carefully studying the pipe-smoking, virile, self-made Brit, dressed, as he always was, in plus-fours as the two squared off, each trying to get a read on the other. Humason, a friend and fan of Shapley, Merrill and Adams was searching for signs that the observatory's celebrity scientist wasn't looking for undue sympathy. Hubble was no doubt trying to get a handle on just what kind of guy Humason was as well. Would he try to steal his thunder or undermine his work and just how ambitious was he, anyway?

In January, Humason had published a paper with Seth Nicholson on his discovery of a massive binary star system (H.D. 163181). Nicholson, a master at determining orbit, was happy to pitch into help his old friend with the project. Neighbors for years, Nicholson had served as 15-year-old Billy's Boy Scout master, and Milt and he had recently bought the rights to a cabin in a small clearing near the West Fork of the San Gabriel River where they planned to make many a fishing trip in the coming months and years.

In the end Humason's credentials and his easygoing manner along with Adam's endorsement convinced Hubble that it was safe to disclose his ideas to him. In a conversation that lasted only a few minutes Humason learned of Hubble's discussions at the conference in Leiden on the theoretical relationship between distance and velocity and of Lemaitre's interpretation of Einstein's general relativity equations. If the relationship could be verified it would corroborate Lemaitre's theory that the entire universe was born of a single cataclysmic event. Hubble was "rather excited," Humason later remembered, and asked "if [Humason] would try to check that out." Humason agreed. Little did he know that after years of struggle, and sometimes frustrated, attempts to establish himself as a serious researcher, he was about to step out onto the world stage.

* * * * * * * * * * * * * * * * * **

Previously, the fastest known redshift for a nebula was 1800 km/s, recorded in 1921 by Lowell Observatory director Vesto Slipher. In 1915, while Einstein was rolling out his revolutionary theory of gravity, Slipher presented a study of fifteen galaxies at the American Astronomical Society's August meeting, eleven of which he reported were redshifted and so apparently moving away from us. Hubble had been studying nebulae for years and had distance measurements (using Cepheid variables as markers) to twenty or more. He had measured spectra for a few of these and found they were fitting neatly into a line of increasing velocity relative to distance. The trouble was that as the distance to these galaxies increased they were getting too faint for Hubble to get steady spectra from them. Hubble needed Humason to delve deeper into space and hunt down these fainter nebulae so that he could determine whether the relationship between velocity and distance held up.

To make life as easy as possible on himself, Humason chose a couple of the brighter nebulae in the Pegasus Group, a cluster of galaxies in the constellation Pegasus. Of these NGC 7619 was deemed to be farthest away based on measurements to Cepheids resolved within the galaxy. On the day he was to start his observing run, Milt phoned up the mountain to ask Tom Nelson, the night time assistant on the 100-inch telescope, to attach the Cassegrain focus cage on the front of the tube and then fit the telescope with the Cassegrain spectrograph VI, the latest in a line of improved spectrographs designed for deep space research. The spectrograph had a 24-inch collimating lens, two prisms and a 3-inch camera that housed the plate Humason would use to record the galaxy's spectrum.

After years of hiking or riding to the top of Mount Wilson, Milt had finally managed to buy his first automobile, a 1928 Studebaker, GB Commander Big Six.

Fig. 7.2 Milt's first car, a 1927 Studebaker Commander Big Six. Milt sits in the driver's seat with Billy to his right and Helen in the back seat

With a 6-liter, 354 cubic-inch engine the Commander Big Six was one of the fastest and most durable cars built at the time. In October 1928 the Commander set the 15,000 mile speed record, averaging over 65 miles/h. In the years to come his penchant for speed would earn Humason the nickname "Mile a Minute Milt," a name given him by his often terrified passengers. Apart from his need for speed, Humason no doubt saw a practical application in buying the burly speed wagon, which would have no difficulty transporting him and his entire family up the rugged mountain road. After saying goodbye to Helen, Milt dropped by the office on Santa Barbara Street to wrap up a few odds and ends and then drove up to the observatory to drop his things in his room at the Monastery before beginning his run (Fig. 7.2).

Modern stargazers have little understanding of how difficult a task charting and chasing the stars was at the time Humason was doing his work on deep space nebulae. In the age of computer technology an astronomer need only punch in a few coordinates and watch from the comfort of his or her chair as the telescope moves into position and guides along effortlessly, capturing the sought-after data digitally. In 1928 the operation of the telescope, the dome and the camera were all done manually, always in near total darkness and often 30 feet. above the concrete floor of the observing platform. At that height one misstep could mean disaster, as Alfred Joy would later illustrate in terrifying detail.

Exposure times for stars were often not more than a few minutes. In his previous paper on M101 he had needed as much as 6 h to get a good spectrum. He was expecting to have to go well past that now, just how much he wasn't sure. The seeing was poor that week, and Milt struggled to keep the guide star fixed in the crosshairs. The hours ran into days as he stood, sat, laid prone and contorted himself into every position he could think of to keep the telescope fixed on the target, waiting and watching for it through clouded patches of sky. The night air was cool, and Humason struggled to stay awake as the hours dragged on, periodically closing the shutter to allow the mercury foaming up in the bearing houses to settle. His feet could get so cold his toes would ache at times, and there was always the danger of falling asleep and toppling from the platform. Waking from one of these dream states to catch himself before he fell was an eye-opener that gave him some much needed adrenaline but did little to warm his frozen fingers. After hours that ran into days, Milt finally closed the shutter on his first attempt at NGC 7619. The next morning Humason exposed the plate and found the spectrum to be unreadable. Almost 20 h of exposure time and more than twice that in time spent squinting at the eyepiece had been in vain. Humason phoned Hubble in Pasadena and gave him the bad news. Hubble, trying his best to shield his disappointment, was supportive in his reply. With still several days remaining on this observing run, Humason resolved to try again and spent the day reclining, half asleep, half in a dazed state of resolution as he tried to predict the exposure time for his target.

That night Humason returned to the dark hole in the sky he had fished for so many hours and cast the telescope's big eye back into the void, peering into the still and starry blackness as he gently and skillfully pivoted the telescope and dome, tracing the line NGC 7619 was known to cut along the horizon. At around midnight, Nelson took his place, and Humason made his way to the midnight lunch shack for a cup of coffee, some bread and a hard-boiled egg. As he closed the door to the Hooker Telescope dome, he pulled a wad of tobacco from the pouch he kept in his back pocket and set it between his cheek and gums.

The night air was crisp as he started for the bridge that led across the arroyo to the lunch shack. Milt pulled a flask from his waistcoat and took a swig of the panther juice he liked to carry with him and glanced up at the night sky. The seeing was better tonight than it had been. If the weather held he might ultimately snag his quarry. His mind drifted to the cabin in the woods where he and his friend Seth Nicholson spent weekends fly fishing the West Fork of the San Gabriel River. Hugo Benioff made the trip when he could, and recently Walter Adams even joined them for a weekend in the forested splendor of his home away from home. Of course Helen and Billy would come. Billy, who had recently become the youngest eagle scout in the history of the Boy Scouts of America, was getting to be an expert fly fisher in his own right. Helen was the rock of the family and no stranger to mountain living. She kept an excellent camp and never shied from adventure. Milt's attention was brought back to Earth by a raccoon and her kits sniffing at the door of the lunch shack. She knew Milt well and rounded the kits up near the door to wait while he fetched her some bread.

After his half-hour reprieve, Milt returned to the dome where Nelson had the telescope ready to begin the second leg of the night's journey. Weary but unbroken, Milt settled into the wooden chair behind the telescope, easing it into exactly the right position and opening the shutter to begin his exposure again. He worked for three more hours or so until, as the distant galaxy began to set in the horizon, Humason closed the shutter and placed it in the dark room on the observatory floor for safe keeping. It had been a good night, just over 8 h of exposure time. If the sky stayed clear for the next three days he might bag a spectrum of the far-off galaxy.

Humason returned for three more nights, but the weather was spotty at best. On the fourth night the seeing was good again, and Milt seized the opportunity to go for more hours of exposure time than he had previously planned. By the end of his second run he had a 33-h exposure of the spectrum of NGC 7619. A bleary-eyed Humason decided to wait until morning to resolve the negative. He pulled the plate and set it aside while Nelson set the telescope into its resting position and closed the dome shutter. Weary from his long observing run, Humason staggered to his room and collapsed on the bed.

The next day Humason arrived at the Carnegie Institution of Washington offices in Pasadena and strode to Hubble's office. Hubble was eagerly awaiting word of Humason's second run and opened the door with the look of a schoolboy waiting for a girl to accept his invitation to the prom. Humason showed him the plate and the negative and explained that he had asked Miss McCormack, a "computer" at the observatory, to take a measurement of the velocity so they could compare figures. The mean value between the two gave a velocity of +3828 km/s, more than twice the velocity of Slipher's earlier velocity for NGC 584. An expression of joy and excitement washed over the self-styled English expatriate's face as he sat at his desk to mark the point where the new galaxy sat on his velocity-distance graph. Bingo! The galaxy landed right at the top of the line with the other galaxies they had been measuring. Adams came into see what all the fuss was about, and in a few minutes the entire office was abuzz over Humason's new velocity.

However, Humason wasn't that impressed by the quality of the spectrum he had taken. The weak lines were barely acceptable for measurement although they did clearly indicate a large shift toward the red side of the spectrum from any plate he had previously measured. Humason knew he needed another exposure, longer and steadier if he was to be sure of his measurement. That was not a job he was eager to undertake. His partner, Hubble, seemed to be blissfully unaware of the hardship Humason had faced in gathering his new evidence and began talking about pressing on and gathering new velocities for still farther off galaxies. Humason agreed to work on anything within the realm they had already measured and said that he would make another attempt at NGC 7619 in due time. In the meantime, maybe they could make some adjustments to the camera to find a way to reduce the exposure times to fainter objects.

Once he had mustered his courage and stamina, Humason returned to the mountain to try again for a good spectrum of the distant galaxy with its extreme velocity. This time he figured he would try for 30 % more exposure time in the hopes of getting a good clear image of the spectrum. For an entire week he sat at the telescope focused on one fixed point in the sky, sliding along with it set right between the crosshairs of the eyepiece, wrestling the elements, through cloud cover and the tumultuous mercury floats that periodically stopped a run while he and Nelson swept up the mercury that had leaked from the bearing houses and poured it back in.

After 45 h of exposure time Humason pulled the plate and plodded over to the dark room on the mountain to resolve the plate into its spectrum. He was pleased to find that his efforts had paid off; there before him sat a clear spectrum for the galaxy, one that would give a much more accurate measure of its velocity. After measuring the spectrum and comparing notes with Miss McCormack he came up with a mean value of +3754 km/s, to which he gave double weight due to its superior nature. The weighted mean velocity for the galaxy Humason estimated to be +3779 km/s, which was in good agreement with the first measurement but with a much more reliable plate this time.

Satisfied that they had the evidence they needed to promote Hubble's new observed relationship between velocity and distance, the two decided to publish their findings separately in the proceedings of the National Academy of Sciences. The two papers were communicated to the body of the academy on January 17, 1929. In his one-page paper, entitled "The Large Radial Velocity of NGC 7619," Humason laid out the stellar coordinates of the galaxy and wrote briefly about the region of the sky where it sits and of Hubble's plan to see if a relationship between velocity and distance could be observed. He mentioned Slipher's previous velocity for NGC 584 and hinted at Hubble's velocity to distance relationship that followed immediately after his own. Hubble's paper, "A Relation Between Distance and Radial Velocity Among Extra-Galactic Nebulae," exposed the world for the first time to the notion that the universe was expanding. Humason had measured velocities for twenty-four galaxies while Hubble measured their distances using Cepheid variables as a guide and charted them. Hubble calculated that for every 3.26 million light years an object was from Earth it was receding at an additional 500 km/s and introduced the equation $v = Kd$ to describe this relationship. The results were in good agreement of the proposed linear relationship. Hubble had his second major breakthrough in five years and now he had a worthy counterpart to help him gather the evidence to advance his work. In the words of the astronomer Allan Sandage years later: "This was the most fundamental discovery in cosmology for all time. It is the foundation upon which cosmology rests."

* * * * * * * * * * * * * * * * * **

As news of the apparent expansion of the universe spread, Hubble became a media darling. Reporters from all over came to his office in Pasadena to get his take on all manner of scientific phenomena. A month after his paper describing his law governing the relationship between distance and velocity was read before the body at the National Academy of Sciences, Hubble wrote a column in *The Times* on Einstein's theories, to which the great theoretical physicist was adding his version of a unified field theory combining gravity and electromagnetism. (Einstein's theory failed under scientific scrutiny. No unified theory has yet been established.) The press was creating a pitch battle between Einstein's static universe and Hubble's expanding one. The battle pitted the world's greatest scientist whose wit and charm made him a favorite with reporters, and the upstart astronomer turned celebrity scientist, with his stoic countenance and the air of an English gentleman.

So which idea was correct? Neither Hubble nor Humason knew for sure that their new expanding world was a reality. Hubble had made his calculations based on the evidence he could collect at the time, and the need to go to print was due in no small part to his desire to be the first to report on what was, if it was true, a historic discovery. Hubble and Humason both knew that more work would have to be done, and that it would require Humason to reach deeper into space in search of spectra of fainter and fainter nebulae.

* * * * * * * * * * * * * * * * **

The night of his discovery Milt had a nightmare that he later described to Mayall. After finishing up at the telescope he walked out of the dome. The pale light of dawn loomed overhead as he walked down the path toward the Monastery. Suddenly he heard the clatter of the slit motor and turned to see the slit door beginning to rise to its open position like a garage door. That was clearly the wrong direction. The dome doors should be closing at this hour. The door stopped short half way open and looked as if it was jammed. The predawn air was still, and the trees began to close in around him as the door began to buck and clack as if it were coming to life. An instant later the bottom of the door raised up, and to his utter amazement threw itself down on the ground in front of him. Afraid for his life, Milt took off down the path as the door slid and slithered after him, gaining on him with every step he took. As he approached the Monastery the giant door closed the distance between them, rattling like a giant metal snake. At the Monastery there was nowhere to hide, and the great metal beast was already upon him, bearing down and preparing to crush him against the Monastery wall.

With that Milt opened his eyes to find he was staring up at the bedroom ceiling. Sweating profusely he turned to find Helen was still fast asleep. He was glad he hadn't bothered her with such foolishness. As he lay in bed remembering his dream he thought of the conversation with Hubble the afternoon before. It had been an incredible journey, and Hubble was certainly keen to continue it. The work had nearly killed him. Despite Hubble's excitement and his own will to succeed at every level nothing could persuade him to return to the work with Hubble. Deeper space? Fainter nebulae? Not before a suitable replacement for the short-focus camera could be developed. In the face of Hubble's urgent pleas and under overwhelming pressure to continue, Humason walked away from the project.

In an interview later in his life, Humason remarked that getting the spectrum of NGC 7619, "…was not a very happy kind of work." This was surely the comment of a man for whom the experience of that year had been softened by time.

* * * * * * * * * * * * * * * * **

Perhaps the perfect antidote to his problem with Hubble was the arrival of his sister from Holland with her 5-year-old son, Willem, Jr. The two had arrived from Rotterdam aboard the cruise ship *Veendam* in October and had been visiting with family and friends in the area. Lewis had resigned his position at the observatory January 1, 1928, and taken a job as a technician at the University of California at Los Angeles. Although she had put down as her regular work as what she did at the observatory Beatrice continued to work for the UC computing department on a part time basis.

The site of Ginny and her young son was cause for celebration. The Roaring Twenties were at a peak, and the world's greatest cities were experiencing sustained economic prosperity. In music clubs everywhere flappers danced to live jazz music in skirts that would make their mothers blush. The movie industry was booming, the automobile was ubiquitous on the newly paved streets of modern cities and giant baseball parks and other sports facilities were being erected by increasingly wealthy businessmen eager to create ever greater moneymaking opportunities for themselves and their board members. Families gathered around radios in their homes to hear a wide array of sports, music and entertainment programming. Charles Lindbergh had successfully flown solo across the Atlantic Ocean in 1927, landing in Paris on May 21 to a cheering crowd of amazed fans. Taller buildings were being built in America's biggest cities, where populations were steadily on the rise. Prohibition helped usher in the era of the gangster, as Al Capone, Lucky Luciano and other mobsters made fortunes in the production and sale of alcohol. Ginny and her son stayed for several months before returning home to Holland. She wouldn't return for nine years.

* * * * * * * * * * * * * * * * * **

The news of the gift of funds by the International Education Board in October of 1928 for the creation of a 200-inch telescope to be built in cooperation with the Carnegie Institution of Washington caused a lot of excitement around Mount Wilson. The telescope was once again the brainchild of George Hale, who had dreamed up the idea of creating an even larger light-gathering instrument before the 100-inch Hooker Telescope was even finished. Early designs were for a telescope of 300 inches aperture, but this plan was soon abandoned as being unreasonably grand in its scope. As part of the plan to find a suitable location for the proposed telescope Humason and Ellerman were sent out to survey sites in southern California during the year.

Desperate to continue his onslaught of deep space nebulae, Hubble went to Adams for help. As one of Humason's early champions, Hubble figured Adams might be able to reach the reticent master spectroscopist. Understanding the implications of Hubble's (and Humason's) work to the advancement of science Adams asked Humason to come by his house one afternoon for a visit and to view his champion hares. Adams was raising rabbits in the backyard of the home on Hill Avenue that he shared with his wife, Adeline, and their two sons, Edmund and John. (Adams's first wife, Lilliam, died in 1920. He married Adeline Miller in 1922.) Milt listened to the observatory director intently as the latter made his case for carrying on the work in the name of science and posterity, but confided he was more interested in Adams' rabbits than he was any conversation that involved going back into Hubble's project under the circumstances. The equipment currently on hand was simply inadequate to insure good results, and the work was too tedious and too dangerous to undertake, especially with no return on his investment of time, life and/or limb. As an expert spectroscopist in his own right, Adams could definitely identify with his younger, more skilled counterpart.

After the two men parted ways that afternoon Adams knew he would have to pull out a wild card if he was going to have any luck convincing Humason to go back to the work he had begun with Hubble. To help make the case Adams decided to call George Ellery Hale to see if he could influence the situation.

One day shortly thereafter Milt got a call from Hale, who invited him out to his solar laboratory at his home in Pasadena. Now 60 years old, Hale spent many of his retirement years away from what he loved most, solar research, only dabbling in it as his health permitted. His mind was still sharp, and he had built a fine solar lab with a small dome and an improved version of his spectroheliograph in the back corner of the property.

Awed by the invitation from the great man of science and founder of the observatory where he had spent so much of his life, Humason drove out to Hale's house and was ushered in by the honorary director's wife, Evelina. Hale took him into his library and the two talked for a while. The stock market had crashed, and the effects on the U. S. economy were reverberating around the world. Businesses everywhere were failing, international trade was plunging and unemployment was on the rise.

Humason filled Hale in on efforts to tighten belts around the observatory as well as the latest developments in research, including the work he had been up to with Hubble. Hale listened intently, enthusiastically adding a tale or two from the early days on the mountain. In time the conversation led them to the problems confronting Humason on the 100-inch telescope and spectrograph. Hale applauded Humason's drive and skill and admitted that he had been all wrong in resisting promoting Humason to the staff when Adams first urged him to do so years before. Hale promised a faster camera with an improved lens if that was what was needed for Humason to continue his important work. Humason agreed the work was necessary and maybe fundamental to the science of universal evolution, and told the honorary director that, if such a camera were to be made available, he would redouble his efforts for Hubble's program.

To aid in the creation of the new camera and a spectrograph that could be used with it, Hale solicited the genius advice of John Anderson, the chief executive officer of the 200-inch telescope project, which was in the design process at the physical laboratory in the basement of the Carnegie offices in Pasadena. Anderson was the head of the new telescope project being financed by John D. Rockefeller and an expert in optics and engineering who had been with the observatory on a permanent basis since 1916. Since, in Hale's view, the new camera would likely be used in conjunction with the new (and still bigger) telescope, it made sense to have the engineer of the telescope design a camera to be used with it.

* * * * * * * * * * * * * * * * * **

A terribly dry spring and summer gave way to a very wet winter on Mount Wilson. Total precipitation between June and January 7, 1930, had amounted to only 0.3 inches. A massive storm on January 7 dumped 55 inches of snow on the mountain, and before the month was out over 100 inches would be recorded, the second highest total in the observatory's history. Fortunately for everyone living and working there, Dowd had recently replaced the two large batteries that backed up the power lines in case of a power outage.

On February 18, 1930 the planet Pluto was discovered by a young astronomer named Clyde Tombaugh at the Lowell Observatory at Flagstaff, Arizona. In 1929, Vesto Slipher, reinvigorated by the notion of discovering Pluto, had set Tombaugh to the task of locating it, and less than a year later Tombaugh had delivered. When news of the discovery got out Nicholson set out to find the planet on observatory plates and determine its orbit. Two of the plates, it turned out, had been made by Humason during his research in 1919. Milt's misadventures in the search for the elusive planet were well documented in the observatory's records that year. Nicholson's own account published the same year detailed the discovery of Pluto from Humason's plates:

> Negative copies, showing images of Pluto, made to scale from two photographs of 2 h exposure by M. L. Humason with the 10-inch refractor of the Mount Wilson Observatory, Carnegie Institution of Washington. These images were found in 1930 after Pluto had been discovered at the Lowell Observatory and a preliminary orbit computed. The planet is located by two lines marked on the plates. The position of "planet O" published by W. H. Pickering in H. A. 82, 1919, is marked by X on the photograph of December 29, 1919.

Several months after his discussion with Humason, Hale could report that Anderson had a new camera ready with a lens provided by Dr. W.B. Rayton of the Bausch and Lomb Optical Company. Humason wrote about it in the Contributions of the Carnegie Institution of Washington Yearbook in May of 1930. The camera had a 1.3-inch aperture, almost a third the size of the one currently on Mount Wilson, and housed a plate that was a mere 1 5/16 inches × 5/8 inches in size. In his words the extremely short focal length would "reduce(s) the spectrum" of deep space nebulae, "to the smallest possible length and width, thereby cutting down on exposure time." Unlike previous attempts to create such a camera, this one preserved the Fraunhofer lines Humason and other spectroscopists needed to carry out their measurements, in good definition.

To test the new camera Humason attached it to the spectrograph on the 100-inch telescope and aimed it at the same point in the sky that had so tried his patience a year earlier. He got the same result, only this time he needed only 12 h of exposure time. Humason knew he now had one of the crucial elements he needed. With an improved spectrograph, designed to work directly with the Rayton camera lens, he could then press forward in the nebular research for Hubble's program.

In a matter of weeks the new spectrograph was ready, and Humason prepared himself to go back to work on nebular research. Finding these incredibly faint objects was difficult even for an instrument as powerful as the 100-inch telescope. To find a nebula deep enough in space to produce a spectrum with a redshift along the lines that Hubble was predicting Humason would have to go after remote nebulae in clusters whose distances could be determined from stars resolved within them. Once he had a predictable outcome he could then get a spectrum for a nebula within these clusters.

Over the course of the next year Humason photographed the spectra of 46 galaxies in seven clusters. These were accompanied by direct photographs of the regions where the nebulae were located. With every new measurement, Humason was peering deeper into space than any human had ever gone.

During a run at the 60-inch telescope, William H. Christie had made a direct photograph of a cluster of several hundred nebulae in the Leo constellation at a distance of about 100 million light-years. Within this hornet's nest of galaxies about five stars in twenty-five nebulae could be seen. This was the farthest known cluster of nebulae discovered at the time. Christie had become a recent recruit of Humason, who was busy trying to find worthy observers to take on some of his other projects. The nebular research was too time consuming to think about continuing his other work with Adams and Joy, Merrill and others to the extent he had been in the past.

Humason spent Valentine's Day at home with Helen and then packed his bag and headed for Mount Wilson on Sunday, February 15, 1931. When he arrived at the observatory he checked in with Nelson, who had attached the new spectrograph and camera to the telescope at the Cassegrain focus and opened the shutter doors to allow the air temperature inside the dome to equalize to the outside air. The weather was lousy, but Milt figured he'd give the cluster in Leo a shot that night anyway.

An hour passed, and it looked like the cloud cover wouldn't let up. Milt and Nelson decided to head over to the 60-inch dome to see how things were going over there. A new kid named Nicholas "Nick" Mayall was over there on a run to get spectra for stars in the Kapteyn Selected Areas. Long on the list of Adams and Joy for their program to catalog stars in the Kapteyn Areas, these stars appeared in the winter sky and were left for last. Mayall had come down from the University of California at Berkeley to work with Seth Nicholson on orbit calculations but was eager to work with both Hubble and Humason as well. Humason knew he had given him some of his dirty work and wanted to see how he was making out.

Entering the dome he found an enthusiastic Mayall filled with questions about the operation of the telescope and dome and the best way to set up the telescope for a run with the Newtonian cage and tracking during partially cloudy nights. Milt spent some time with Mayall answering his questions and explaining the idiosyncrasies of the telescope's clock mechanism and movement. Checking the weather and seeing the cloud cover wasn't going to let up for a while, Humason decided they should continue their business at the pool table in the basement of the Monastery. That cluster in Leo would have to wait for another day.

The next three nights brought the same if slightly better seeing on the mountain. Humason stayed at the telescope peering through the eyepiece trying to maintain his focus on the guide star as one wispy cloud after another passed overhead. Just after midnight on the morning of February 19, Humason wandered into the midnight lunch snack to find Mayall seated at the table sipping tea. His young protégé said he had little to show for his efforts and asked if Milt had fared any better. No, was the reply, and after four nights he had only managed a total of 13 1/2 h of exposure time. When Mayall asked how much time he thought he needed Milt shrugged his shoulders and said he didn't know. He was going after a spectrum for an object with a visual magnitude of 15.5, 6300 times fainter than could be seen with the naked eye, using equipment that had been newly developed. He had no idea if the exposure time would be enough. He was doubtful in any case that the definition would be clear enough to submit as evidence for the expansion principle. With the weather only worsening, however, Milt had seen enough for this run. He was going to develop the plate and see what he had. He was headed to the darkroom to develop the plate for the cluster in Leo and wondered if Nick would like to come along to see how it turned out. The young college grad jumped at the chance and accompanied Milt to the darkroom. There in the dim red light of the room Mayall watched as Milt ran the plate through the series of bins and held up the negative wet with hypo to see it for the first time in the light. As he watched, Mayall saw Humason's eyes fill with amazement at what he saw, a clearly defined spectrum with an enormous redshift! Excitement filled the room as Humason showed Mayall the telltale H and K lines in the series that were shifted to the red end of the spectrum by nearly 300 Å. He couldn't believe how well the new camera worked. With this equipment he could observe much fainter nebulae than he had thought.

He would call Hubble in the morning. Now it was time to celebrate. The two men made their way to Humason's room in the Monastery. Mayall, overwhelmed by the experience, was as much in awe of the older man's skill and expertise as he was taken by his easygoing charm. Mayall sat on a chair at Humason's desk as the latter rummaged through some things and dug out a bottle of hooch. "A toast," said Humason as he unscrewed the cap and looking around in vain for a cup to pour the elixir into stretched the bottle over to his companion. "What's this," Mayall asked? Milt grinned, a glint of mischief in his eyes, and said, "Panther pacifier." After Mayall recovered from the taste and potency of the Milt's fiery concoction, the two men sat for a while and talked about the road ahead. It was the beginning of a professional friendship that would last for forty years.

Later that morning, as noon approached, Mayall joined Humason in the lab at the observatory building on Mount Wilson. Milt picked up the phone and dialed Hubble at his office. "Mr. Hubble," he said, "I got a fine plate of the brightest nebula in Christie's cluster in Leo, and it shows just about the redshift you predicted, nearly 20,000 km/s." Hubble thought for a second and replied, "Well, now you are beginning to use the 100-inch the way it should be used."

Later that afternoon, Humason arrived at his office on Santa Barbara Street and was immediately visited by Hubble, who was eager to see the new spectrum. Looking down at his desk, Milt found a letter addressed to him from George Hale. The letter read: "I was delighted to hear of the immense velocity of the small nebula. I must confess that I don't see how you kept this 17th magnitude object on the slit during 13 1/2 h."

After twenty-three years of service to the observatory, Jerry Dowd retired on January 1, 1931. A small celebration was held in his honor on Mount Wilson. The 62-year-old engineer had been an integral part of the development of the facility over the past two decades, and had watched his children and grandchildren grow up on the mountain. Billy, who would turn 18 in October and was planning to attend college at

Caltech, was the Dowds' pride and joy. Jerry and Katherine retired to their ranch in Imperial Valley and would visit the mountain frequently in the years to come.

As the new year was getting under way, Hubble and Humason readied their manuscripts for publication. Forty new velocities were added to the current total measured for the velocity-distance scale. As they had done previously, Humason would prepare a paper to be printed just before the paper on the broader topic of uniform expansion. This time, though, Hubble wanted to publish the bigger paper jointly, insisting on giving Humason due credit for his work on the program. The two had come a long way, and the nod to his partner was a telling step for Hubble, who was prone to distrust his fellow researchers. In Humason, however, had found a respectful and inestimably valuable collaborator. Humason's gift for stellar photography and his pioneering and inventive spirit had earned him the respect that Hubble was not inclined to dole out readily. After all, it was Humason who had photographed all of the spectra and measured the velocities for his program. Hubble knew this and was prepared to acknowledge it (Fig. 7.3).

Among those who were impressed by their work was Albert Einstein, who came to Pasadena in January of 1931 on an invitation from Caltech. While in the area,

Fig. 7.3 Milt, Helen and Bill Humason, circa 1931

Fig. 7.4 A distinguished guest. Albert Einstein stands for a photo during his visit to the observatory. From *Left* to *right* Milton Humason, Edwin Hubble, Charles St. John, Albert Michelson, Einstein, W.W. Campbell and Walter Adams (Observatories of the Carnegie Institution for Science Collection)

Einstein arranged to visit the observatory offices in Pasadena as well as the observatory on Mount Wilson for an evening viewing of the stars. Touring the observatory in the afternoon, Einstein went to the top of the 150-foot solar tower in the "bucket," an amusing ride for the uninitiated, and spent a night viewing the stars through the 100-inch telescope. The next day he visited the offices of the Mount Wilson Observatory on Santa Barbara Street, where Hubble and Humason met with the master of theoretical physics and showed him some of the work illustrating their contention that the expansion appeared to be real. Conceding the argument in light of the evidence against him, Einstein joked, through his interpreter, that his constant had been a cosmological blunder.

Later the group stood for a photo in the library under a large portrait of George Ellery Hale. In the photo Humason stood at left next to Hubble, followed by Charles St. John of the solar department. Albert Michelson, whose measurements of the velocity of light had been so influential to Einstein's work, stood beside the legendary physicist. A specially designed apparatus, consisting of a mile-long tube with pumps attached to it to create a vacuum that would eliminate error in his velocity measurement, was in its final phase of development. Unfortunately,

Michelson died in early May, just after the first tests were run of this extraordinary new instrument. Standing beside Einstein was William Campbell, director of the Lick Observatory. Mount Wilson Observatory director Walter Adams stands beside Campbell in the photo.

It was a proud moment for Humason. Standing for a picture in the observatory library beneath the image of George Ellery Hale with such an esteemed group of colleagues would be a moment he would never forget. The former muleskinner had somehow managed to climb the twisted and winding path up the steep slopes of Mount Wilson and land on top of the science world.

In March the team of Hubble and Humason took the world by storm, publishing their findings in the Contributions of the Carnegie Institution of Washington Yearbook No. 426 and 427. Humason opened in his usual deferential way, citing Hubble in the opening of a ten-page discussion outlining his research and the instrumentation used in the study of the spectra. Humason's paper was the first to advance the concept of galaxy groups based on redshifts. Not wanting to forget those who helped him along the way, Humason ends his note by thanking Tom Nelson and Glenn Moore, the relief night time assistants, for their help during his observing runs.

The next paper, Contribution 427 of the Carnegie Institution Yearbook, would become one of the most influential scientific documents of all time, taking the distance scale out to 105 million light years and a velocity of 19,600 km/s. Although cautiously referring to their velocities as "apparent" in both papers, Hubble and Humason concluded that the scale was predictable out to the limits of the equipment available to them at the time. The questions revealed on the evolution of the universe in 427 would occupy cosmologists for the next seven decades (Fig. 7.4).

Chapter 8
The Great Observatory

Abstract The life of the observatory during the economic depression of the 1930s when millions were without work is vibrant and Milt is the life of the staff. In the years leading up to the Second World War, the observatory acts as a beacon of hope and entertainment for those struggling to get by in hard times.

Sitting in his office in January 1932, Edwin Hubble was being interviewed by a reporter for *The New York Times*. During the interview the reporter asked Hubble to comment on the details of a spectrum his partner, Humason, had taken for a faint nebula. Leaning back in his chair while eyeing the photo, Hubble set his pencil down and filled his pipe with tobacco. He struck a match and lit the pipe as he carefully studied the spectrum, preparing his thoughtful reply. Finally, with the reporter hanging on his every movement, Hubble spoke:

> Do you realize the amount of patience, ingenuity and expert experience that little photograph represents? Besides having to guide the telescope continually, Humason had to control the focus and the comparison spectrum…keep the temperature of the spectrograph exactly right – not for an hour or so, but all night. Sitting in total darkness with his eyes on a slit of dim light little larger than a pinhead, he worked levers and punched buttons night after night for a week without once moving the photographic plate or losing sight of that faintly luminous spot in the sky, and here's the result – a report on conditions in a very remote region of the universe. Talk about the romance of the heavens, why Humason has brought it down to earth.

As the 1930s began the Mount Wilson Observatory had become world renowned as a leader in astronomical research. With the world reeling under economic depression the steady stream of world-turning discoveries seeming to emanate constantly from its halls were welcome relief to a public increasingly hungry for something to cheer about. For astronomers and astrophysicists the world over, Mount Wilson had become the Mecca of astronomy. Much excellent work was being done in all departments, with far-reaching consequences and mutual benefits, but by the early 1930s Edwin Hubble had become the face of the observatory. Milton Humason was its heart and soul.

Throughout the 1930s, when the Mount Wilson Observatory was at its peak of importance and stature, Milt was the life of the party among the staff. Whether he was taking their money at the weekly poker games on the mountain or regaling

© Springer Science+Business Media New York 2016
R.L. Voller, *The Muleskinner and the Stars*,
Springer Biographies, DOI 10.1007/978-1-4939-2880-4_8

them with tales of his exploits at the telescope or wading through the California waterways with his fly rod in hand, everyone wanted to be around Milt. His infectious personality was both easygoing and entertaining. Not that he was a perfect human being necessarily, but he was fun! After Santa Anita Park opened in 1934, the consummate riverboat gambler could often be seen leaving work early on certain afternoons to head to the racetrack.

With the publication of their joint paper, "The Velocity-Distance Relation Among Extra-Galactic Nebulae," Hubble's star climbed even higher. The most famous astronomer since Galileo, Hubble carried himself with an air of intelligent English aristocracy, and he had a flair for public theater. He was always prepared for the press, and the press was usually not far away from him. He made friends with poets, actors and dignitaries and threw lavish parties with his wife, Grace, at their home on Woodstock Avenue in Pasadena. Hubble was not only a master astronomer but he was also a master of ceremonies. He was standoffish with colleagues, most of whom he didn't trust, but open to the public (Fig. 8.1).

Humason was exactly the opposite, content to carry out his work and eager to spend his downtime with his family. Where Hubble shined at the microphone, Humason shied away, uncomfortable in the limelight and declining public comment in the press whenever possible. His confidence at the telescope couldn't make up for his lack of education, he thought, and he usually demurred in scientific argument even if he knew he had an opinion worth hearing. Far from lacking in self-esteem, though, Humason shined at private gatherings, where his old-school country charm and ability to turn a yarn made him a favorite guest for everyone around the observatory and at Caltech.

As the two men continued their work on the expansion problem, their professional and personal bond became stronger. They shared a love of fishing and hiking and Hubble would often join Milt and friends from the observatory at the cabin in the San Gabriel range. Benioff, who was now at the Seismology Lab at Caltech, was a constant presence as well.

Fig. 8.1 Milton L. Humason holding a small prime-focus spectrum plate (Observatories of the Carnegie Institution for Science Collection (COPC 2923))

Hubble trusted Humason probably more than any other man in the field, not only for his expertise but his total deference and respect for order and fairness. In Humason he found a true spectroscopist, a man borne of the art who understood his place and was perfectly at ease in his role. His respect for Humason was evident in his inviting his partner to his home for weekly social sessions on scientific matters with members of the Caltech staff and other visitors. Hubble was Humason's one and only shot at a taste of stardom, and Humason knew it. Not that he was particularly interested, but the taste wasn't bad, either. Hubble, too, had the unique ability to offer breathtaking leaps of intuition and supposition while having the restraint to hold back until his beliefs could be better founded on fact. Both men had the quality of backing up who they were with groundbreaking work in their respective fields. Theirs was, one might say, a match that was written in the stars.

* * * * * * * * * * * * * * * * * * **

Walter Baade joined the staff at Mount Wilson in 1931. Already an expert in the field of photometry and probably the leading authority on nebular research outside Mount Wilson, Baade had visited the observatory in 1926, where he met and worked with both Hubble and Humason. Among his many achievements Baade was the discoverer of a large cluster of galaxies in the constellation Ursa Major in 1928. It was the farthest known galaxy at the time, and Humason later used it in his study of spectra for deep space nebulae in 1931. Born in Schröttinghausen in northwest Germany in 1893 with a congenital hip defect that made him limp, Baade had been working for years at Hamburg Observatory but came to Pasadena with his wife, Muschi, for the weather and, especially, for the seeing.

Baade was an exceptional scientist and an equally exceptional person. In many ways he was the antithesis of Edwin Hubble. Unlike his American counterpart, Baade was a hit at parties, an open-minded conversationalist and a kind spirit who listened as intently as he spoke. He was also a practical joker who loved to gamble and go fishing. So, naturally, he and Milt immediately became friends. Shortly after arriving Baade began working with Humason on the nebular work with Hubble but would gradually drop that work to team with Fritz Zwicky at Caltech on supernova research (Fig. 8.2).

At home the Humasons were celebrating Billy's acceptance to the California Institute of Technology. As Billy prepared to head to college, Milt and Helen were arranging to move from their home in North Hudson Avenue to a quiet little place on Oak Knoll Gardens Drive. The three bedroom, two bathroom home was built in 1922 on land that was once a sheep ranch owned by Henry E. Huntington. The home had a beautiful garden and the quiet neighborhood was close to the Caltech campus where Billy was getting his degree in science as well as the observatory offices in Pasadena.

While the work continued with Hubble on the nebular program Milt published five papers on other projects he'd begun in years past between 1931 and the end of 1932. A report on stars with abnormal color was published with Frederick Seares in

Fig. 8.2 Milt hams it up with Walter Baade on a location scout for the 200-inch telescope

January of 1931, and a lengthy follow-up paper on Class Be stars went to print in March of 1932. This work, which Milt had begun with Paul Merrill, was now being aided by several new members of the staff, including Nick Mayall, who had helped out on another paper the two published in December, 1931. Both Mayall and Hubble helped Milt on the completion of a sprawling paper on faint stars in the Kapteyn Selected Areas, giving time during their runs on the 60-inch telescope to take plates of areas Milt hadn't been able to get during his many runs on the mountain. In turn, Humason took spectra of two novae in the Andromeda Galaxy that Hubble had found in July and September. The cooperation between Hubble, Humason and Mayall foreshadowed their future collaboration in establishing the *Hubble Catalogue of Galaxies.*

After the rush to publication on various projects ended in 1932 Humason again turned his full attention to the nebular work. There was speculation within the field of astrophysics whether the redshifts Humason was measuring were real or due to some otherwise unknown law of nature. Not sure and appreciably unimpressed with the miserably small sample they had made their assertion from, Hubble and Humason decided, to play it safe in the press. In a leaflet published in *The Astrophysical Journal* in July 1931, Humason describes the state of their research: "The [apparent brightness of nebulae] are confidently interpreted as distances but the interpretation of the redshifts as velocities of recession is controversial. For the present we prefer to speak of these as *apparent.* In this sense the velocity-distance relation is one of the two established characteristics of the observable region of the universe as a whole. The other is the…uniform distribution of the nebulae. The two together form the observational basis for theories concerning the structure of the universe."

The duo of Hubble and Humason had tentatively disclosed their discovery of expansion in their earlier paper. What they needed now was more evidence to corroborate the claim. In 1932 help came from Harlow Shapley, of all people. Shapley published a new catalog of bright galaxies compiled with Adelaide Ames at Harvard. The Shapley-Ames Catalog was a useful tool in crosschecking the velocities Humason was measuring and magnitudes, from which Hubble would calculate distance. Their 1934 paper extended the total number of known velocities to ninety, adding thirty-five isolated nebulae to their total at velocities ranging from 300 to 19,600 km/s as measured by Humason in 1931. With the addition of more than three dozen galaxies to their sliding scale of velocity to distance they had gone a long way to providing the evidence that the universe was expanding. It was time for a more pronounced program of nebular research, one for which they would need a partner.

Nick Mayall finished his thesis at the Lick Observatory and was awarded a Ph.D. in science in 1934. Upon graduating the young observer was hoping to get a position on the staff at Mount Wilson. Due to the depression and the recent rash of new hires no positions were open in Pasadena. In view of the circumstances, Mayall decided to stay at Mount Hamilton and take a position at the Lick. While working on his thesis Mayall had designed a spectrograph to be used with the 36-inch Crossley telescope. Now a part of the staff he pushed the observatory director, Robert Aitken, for the money to build the instrument. With the spectrograph attached to the Crossley telescope, Mayall hoped to join forces with Humason at Mount Wilson to help with the nebular program. Aitken agreed, and the spectrograph was built and installed early in 1935.

The plan Humason and Mayall would undertake was to measure redshifts for the 1246 galaxies in the *Shapley-Ames Catalogue of Galaxies*. They would divide the work according to the best candidates for the equipment available at each observatory, with plenty of crossover between them to test for errors. Mayall began his end of the research that year while Humason readied himself for the final installment on the distance scale problem from Mount Wilson.

By the time he published his 1936 report, "The Apparent Radial Velocities of 100 Extra-Galactic Nebulae," Humason had recorded the velocities for 146 galaxies, pushing the velocity range out to 42,000 km/s and the distance to the farthest known nebula at 240 million light-years. Hubble's predictive distance scale was holding up nicely to this point. A uniform expansion for all galaxies tested seemed to be a reality, but to really be sure, they both agreed they would need to push the velocities out to between 60,000 and 100,000 km/s, at which time, Hubble figured, they might reach the visible limit of the universe.

Humason had reached the visible limit of the 100-inch telescope that much was for sure. After years of prodding and poking at the sky, he had exhausted the telescope's capabilities. They decided to shelve the program in preparation for the

Fig. 8.3 Breakfast is served. Milton Humason (at table *right*) shares the table with legendary Mt. Wilson 100-inch assistant, Tom Nelson (facing the camera)

commission of the 200-inch telescope that was in development and scheduled to go online sometime around 1940.

Incredibly faint and distant, and moving away from Earth at an unfathomable rate of speed, Humason's speeding galaxies were a marvel to the general public. At the age of 45 he had become a consummate professional, one of the world's leading authorities on the state of stellar photography, the instruments that powered the observatory on Mount Wilson and on the visible expansion question. Writing in the Astronomical Society of the Pacific journal in 1936, Humason published a leaflet entitled "Is the Universe Expanding?" It was the question on the lips of anyone who was keeping up with the latest developments in science. In fact, it was the fundamental question of the era, and the public was thirsty for answers (Fig. 8.3).

Humason's accounting of the problem reviewed the evidence from Slipher's first measures of the radial velocities to Hubble's discovery of Cepheid variables in Andromeda, which gave the first physical evidence that galaxies existed outside our own, to his own work using the large reflectors at Mount Wilson. He explained the notion of the "apparent" expansion by means of Hubble's formula on the velocity-distance relation and summed up his arguments by stating "the universe… probably stretches on and on, far beyond the limits of our telescopes."

Adding to the specter of the expansion phenomenon the honorary observatory director, George Hale, had commissioned the 200-inch telescope in 1928 on behalf of Caltech, with a $6 million grant from the Rockefeller Foundation. After considerable research to ascertain observing conditions at several sites, it was decided that nearby Mount Palomar provided the best overall location to build Hale's new observatory. Work was beginning on the dome at Mount Palomar about 90 miles south of Pasadena, and a new glass blank for the mirror was being formed with a new technology called Pyrex at the Corning Glass Works in New York. The stage was being set for the final showdown on the expansion principle.

Part III
A New Era (1938–1964)

Nothing is built on stone; all is built on sand, but we must build as if the sand were stone.

Jorge Luis Borges

Chapter 9
The War Years

Abstract As war grips Europe and Asia, the observatory and staff are put into service aid in the war effort. Milt works to protect his family from the ravages of the conflict's effect on the homeland. When Billy's new wife, Ruth, contracts tuberculosis in the early stages of the war, he must leave their young daughter with her grandparents while he tends to his wife's ailment. As the war ends the observatory gets back to completing the Hale Telescope on Mt Palomar.

The year 1938 started auspiciously enough. Late in June of the previous year, Ginny had arrived in Pasadena with her sons, 14-year-old Willem and 11-year-old Lewis. Having once traveled to the area when Willem was younger, Ginny was eager to show her youngest son her childhood home. With the unrest in Europe Ginny and Billy, Sr., decided she had better visit the United States before the violence carried over to the Dutch homeland. She and the children stayed long enough to celebrate Milt and Lewis' birthday. Lewis had divorced Beatrice and was living by himself in Los Angeles, still working at the UCLA. During her stay Ginny implored Milt to come to Holland for a visit so she could show him and Helen her family's beautiful home and the country they lived in. As it happened Milt and Helen were planning to be in Europe the next year for a conference Milt had been invited to, and they were eager to spend some time touring the region before returning home. They all agreed that a visit to the Suermondts was a must during their European tour.

But first things first. Billy, who had graduated from Caltech in 1936 and taken a job teaching at his alma mater, was preparing to marry his sweetheart, Ruth Petty. Ruth had received a degree in nursing from Saint Vincent's Hospital in Los Angeles in 1934 and worked in a Los Angeles hospital. After the wedding, Milt and Helen were headed overseas to visit his sister, Virginia, who was living in Holland with her husband and two sons. Milt was planning to attend the 6th General Assembly of the International Astronomy Union in Stockholm, Sweden, and the two were planning to spend some time traveling through parts of Europe while on their journey. There was sure to be much discussion at the conference over the pending dedication of the 200-inch telescope, sometime around 1940, and Milt looked forward to spending time with friends and family. In the spring he and Nicholson and friends might head out to the cabin for a few days fishing and hiking.

© Springer Science+Business Media New York 2016
R.L. Voller, *The Muleskinner and the Stars*,
Springer Biographies, DOI 10.1007/978-1-4939-2880-4_9

In Europe by that time conditions were worsening, and increasingly the growing specter of unrest was beginning its slow and steady March toward all-out war. In 1935 the Italian dictator Benito Mussolini had invaded Ethiopia, assuming military control of the country in 1936. After joining forces with the German Chancellor Adolph Hitler, Mussolini had thrown his political and military support behind general Francisco Franco of Spain in his effort to overthrow the Spanish Republic.

In Asia things were no better. Japan had invaded China in 1937, capturing the capital, Beijing, in July. The Chinese government relocated and continued the war while the Japanese set their military sights on the Soviet Union, where they were summarily turned back by the Red Army. The ineffectiveness of the Japanese war machine in the area had convinced the Japanese government to adopt a conciliatory strategy with the Soviets while turning their attention toward Europe and the United States.

Altogether things looked bleak on the European and Asian continents. In America the general public, remembering the agonizing loss of life in the First World War, were once again taking an isolationist view of the problems facing the rest of the world. There was enough to deal with at home without having to get involved in the chaos erupting in Europe.

Far closer to home, Milt and the members of the staff at the Mount Wilson Observatory were suffering from tragedies of their own. Francis Pease, designer of the 100 and 200-inch telescopes and one of the original five members of the observatory, died on February 7 after complications from surgery. As the observatory mourned the loss of one its finest and most accomplished members, the staff were greeted by even more tragedy when George Ellery Hale died on February 21, at age 69. Although his failing health prohibited his participation in the ongoing research at the facility that he had given so much of his time, energy and health to create, his spirit had always been present, and he was beloved by all members of the staff. In a brief salute to his longtime friend and mentor in the Contributions of the Carnegie Yearbook that year, Adams wrote that Hale would be "remembered for the many great institutions which he conceived and established, and for the breadth of his outlook upon the progress of science and its part in human life." Adams noted Hale's "constant encouragement of his associates to undertake investigations freely and independently with every facility he could afford them," adding that "no one could be associated with him without acquiring a deep personal affection for him."

In just two weeks, the world had lost two great men of science of the twentieth century. As if nature were lamenting the deaths a massive five-day storm moved into Los Angeles in late February, dumping rain on the area. In what still is one of the worst natural disasters in the city's history, rainfall unleashed major flooding, damaging or destroying over 7000 homes and killing over a 100 people. Milt was on the mountain when the storm hit and poured twelve inches of rain on the observatory in one 42-hour-period. Forty mile per hour winds nearly tore the roof off the solar observatory, but workmen tied it down. Later the rain gave way to snow as the entire mountain froze over in an icy bog. Landslides washed out the road to the observatory, which had been paved in 1936, and threatened villages at the base of the mountain. Milt and family were spared significant loss of property and life at home, but in a personal blow to him and his fishing buddies the West Fork of the San Gabriel River overflowed and washed out the cabin they had shared for so many years. To the superstitious Humason, who believed he could locate water using a switch as a divining rod, the event augured bad signs of things to come.

In March the German army marched into Austria and Czechoslovakia and annexed them with little resistance from the rest of Europe. This spurred the international community to act. Receiving word from Washington of the need to begin beefing up U.S. military stores, Walter Adams ordered work on astronomical equipment at the labs in Pasadena to be stopped so that it could be retooled for work on optics for gunsights and other accessories for the armed forces. Slowly, painfully, the increased trials unfolding on the European landscape began to take hold at home and at the observatory.

By the time they published their 1938 paper on the subject of supernovae, Walter Baade and Fritz Zwicky had severed ties with each other. Once friends, Baade had found Zwicky's irascible nature too difficult to handle. Since Hitler's rise to power began in the mid-1930s Zwicky had also become increasingly suspicious of Baade, who had never bothered to get his citizenship papers. For Baade, a return to his German homeland had always been his goal. Succumbing to paranoia, Zwicky became suspicious of the older man's nationalistic leanings, accusing his one-time friend and scientific partner of being a Nazi. The accusation surprised and unnerved the German astrophysicist, who began to distance himself from Zwicky thereafter.

Bill and Ruth were married in a small ceremony in Los Angeles on June 10, 1938. After the wedding Milt and Helen readied themselves for their trip overseas, leaving on a cross-country train in July. After sailing to France they toured Paris and made their way north into Belgium before boarding a train for Holland and a visit with Virginia and her family. Together for the first time since she left Pasadena for Holland with her husband years earlier, Milt and Ginny talked for hours catching up on lost time. They stayed a few days, absorbing all they could about life in the Dutch countryside before it was time to board a ferry for Stockholm and the meeting (Fig. 9.1).

For Milt, who had just become a member of Commission 28 (on Extragalactic Nebulae), the trip marked the first time he'd been invited to the prestigious meeting held and hosted by a different country once every three years. The Swedish government opened the doors of parliament to house the IAU meeting, which opened on August 3, presided over by the president of the union, M. Esclangon and vice president, Walter Adams. Delegates from twenty-eight countries were on hand for the event.

Humason met Adams, Walter Baade, Paul Merrill and Nick Mayall in the morning and the general assembly began at 2:30 p.m. Milt had a chance to catch up with his old mentor, Harlow Shapley, who was there with the Harvard Observatory contingent. Other Ivy League delegates included Henry Norris Russell, whose love of the Mount Wilson Observatory's instruments had been tempered by his distaste for the quality of the water so many years ago. At 8 o'clock Milt and Helen joined other delegates and their guests for dinner and dancing at City Hall. A reception was held on the second day at the Royal Palace held by Prince Gustafus Adolphus

Fig. 9.1 Milt and Helen on a
visit to Virginia Humason
Suermondt's family in
Holland, 1938

of Sweden. Later the delegation was transported by train to Stockholm Observatory for a tour of the facility led by the observatory director, Bertl Lindblad. On Saturday, Paul Merrill gave a dissertation before the combined commissions on "Emission Lines in Stellar Spectra," and later that evening Humason took part in a round table discussion of galactic structure for which Shapley opened with a talk on "Absorption Phenomena as Inferred from the Distribution of Extragalactic Objects." Throughout the seven-day event Milt found he was in high demand as delegates from all over the world questioned him about everything from the status of the expansion question to developments with the Mount Wilson and Palomar telescopes to new coating techniques, optics and the Rayton camera lens.

The day after the convention Milt and Helen took the ferry to the coast of Scotland and toured the country on their way south to Liverpool in England, where they set sail again for home on the ship, U.S.S. *Franconia.* It was August 19, 1938, Milt's forty-seventh birthday and this one surpassed any he could remember. After ten days at sea they landed in New York City and spent two days touring the city before boarding a train for home.

Milt returned to work in mid-September and found among the various papers and letters on his desk a letter dated August 13. It was written by a young Viennese medical student named Alice Grosz who pleaded with Milt to arrange her safe passage to the United States. Her family had been forced to leave the country to avoid persecution, and young Alice had stayed behind to tend to her grandmother. The letter and accompanying photo depict the despair so many felt at the time, as their worlds were being turned on end by Nazi aggression. No application for the girl's passage across the Atlantic survives, but the fact that this letter survived in Humason's archive is illustrative. If he hadn't able to assist Ms. Grosz in escaping her ordeal Milt would certainly have felt the sting of helplessness and frustration so many felt at the time as the country was increasingly drawn into the chaos of the war in Europe.

The heightened threat of invasion and the brazen assaults on sovereign countries in Europe and Asia led to the mobilization of armed forces in America, sometimes with bizarre and deadly consequences. With the majority of the senior observers away, an Airforce A-17 attack plane slammed into the main building at the Lick Observatory, killing the pilot and co-pilot instantly on May 21, 1939. The plane had been flying through clouded skies that shrouded the observatory. Fortunately for Mayall and the rest of the staff at the Lick none of the staff were injured in the crash. Several alert members of the junior staff worked in the moments following the crash to prevent explosion and fire that would have threatened the buildings and instruments.

On March 3, 1940, a massive crowd of 78,000 fans jammed Santa Anita Racetrack near Los Angeles, California, to watch Seabiscuit try for the third time to win the Santa Anita Handicap. The little underdog had captured the hopes of the nation during the depression era, defeating the heavily favored War Admiral in a match race at Pimlico Race Course in Baltimore, Maryland, in 1938. In a career that lasted seven years, the undersized underdog had won every major race except the "hundred grander" at Santa Anita, so named for its $100,000 minimum purse. Milt and his friends Walter Adams, Paul Merrill, Walter Baade, Adrian van Maanen and Rudolph Minkowski (who joined the nebular photography department in 1935), were frequent visitors to the track and were amongst those racing fans who were rooting for Seabiscuit to pull off another miracle. Coming back after an injury in 1939 the tiny horse had lost the first two races on its comeback tour but was starting to find his form again, and a hopeful public had their money on him again.

There wasn't much to cheer about by the turn of the new decade with an unemployment rate hovering around 15 %. To make matters worse, Germany had invaded Poland in September, and two days later England and France declared war on the Reich. Seizing the opportunity the expansionist Soviet Union invaded Poland a week later and further into the Baltic countries in November. The specter of war in Europe was imminent and, although America maintained an isolationist stance, President Franklin D. Roosevelt asked for and received an amendment to the Neutrality Act of 1935 that made it legal to sell arms and munitions to belligerent nations on a cash-and-carry basis, opening the door to a new era of arms manufacturing. In spite of its political neutrality internationally, the American war machine was under construction. This was supposed to have positive effects on the economy and job creation, but no momentum had been felt on the ground.

Like a real life superhero, always seeming to answer the bell, Seabiscuit won the Santa Anita Handicap that day, surging between two horses as they came down the final third to beat his training partner, Kayak II, by a length and half and setting a new course record. Seabiscuit became the winningest horse in history and was named horse of the year for 1940. For Milt and his friends it was a day to remember.

In May the German Blitzkrieg overran military forces in the Netherlands and Belgium. The event deeply troubled Milt, who feared for Ginny's safety and the well-being of her family. In a letter from her shortly thereafter, Ginny wrote of an underground network that had been established to get mail and provisions to and from the country. For their part, Ginny and her husband Billy were sheltering their Jewish friends inside the walls of their offices in Rotterdam. With the town heavily guarded by German and Axis forces, this was a deadly game, and Milt and Helen knew it. They began sending food and provisions frequently, and Helen wrote words of encouragement to try to bolster her sister-in-law's spirits.

Germany, Italy and Japan officially formed the Axis Powers in September, forcing the United States to change its policy to protect the distribution of munitions to the Allies in the Atlantic Ocean. The escort of British convoys by American ships put both in jeopardy of being sunk by German submarines. Increasingly, the staff at Mount Wilson began heading off to different parts of the country to work on the war effort as part of a hidden campaign to aid England in fending off the German army.

Early on Sunday morning, December 7, 1941, 23-year-old Don Nicholson and a friend left Pasadena for a day of surfing. The two friends told jokes and listened to the radio as they drove toward the coast. It was a warm day in southern California, and Don, the son of Seth Nicholson, knew the steady wind blowing in from the West meant good waves were waiting for them at the beach. At a little after 10 a.m. PST a news report interrupted the song they were listening to with a special announcement. The details were muddled, but the essence of the news was as clear as it was bleak. An air raid by Japanese air force bombers at Pearl Harbor on the island of Oahu in Hawaii had devastated the American Pacific fleet. The young men listened to the report and were shocked by what they heard. Outside everything looked the same, the Sun was rising in the East and there wasn't a cloud in the sky. Not seeing any present danger young Don and his friend decided to go surfing anyway.

Later as word spread, photos filled the front pages of every newspaper in the country, and movie houses began showing moving pictures of the carnage unleashed on American soil. Americans such as Don Nicholson (who enlisted in the army) joined in an effort to subdue the Japanese and German empires under the combined weight of Allied military strength.

With the fear of an invasion gripping the nation, air raid sirens sent citizens scrambling for shelter. Billy's wife, Ruth, was pregnant with the couple's first child, and the sound of the sirens sent Billy into the hallway barricading them in doorways or under mattresses. A brownout was enforced in the evening, leaving the city eerily dark. Overnight fellow citizens turned on their Japanese-American neighbors. Internment camps sprang up all over America. In Los Angeles, Santa Anita Racetrack became the temporary home for thousands of Japanese-American citizens who had lived and worked in the country all their lives, now coldly and cruelly spurned by their countrymen, rounded up and sent to the grounds of the race course for the duration of the war.

The national mood was less spiteful toward German-Americans until Hitler declared war on the United States on December 11. Over 11,000 German enemy aliens were detained and sent to internment camps in the center of the country (around 110,000 Japanese were detained) as the war department began evicting people of both German and Japanese descent from coastal areas. Fear of an invasion by sea was high, and coastal cities were on a curfew, with lights out after sunset.

Any German-born citizen who could not produce the proper naturalization documents was being detained.

Walter Baade had been living and working in the United States since 1931 and had never fully applied for citizenship. This was due entirely to his feeling that he would return home to Hamburg to assume the directorship of the observatory there once his work at Mount Wilson was completed. As world war erupted, however, Baade found himself at the center of an effort to remove him and his countrymen from their homes.

Although he was not at all a supporter of the Nazi movement Baade had signed a document that maintained his allegiance to his home country some years earlier. Fearing he would be seen as a traitor and not knowing what would become of his friends and family, he reluctantly pledged allegiance to the German chancellor and leader of the Nazi Party by signing the affidavit "Heil Hitler."

Walter Adams had a different agenda. The aging director was never a fan of Hubble and wanted to keep Baade around, believing Baade was a better pure scientist than Hubble and equal if not better at synthesizing data into plausible solutions. Keeping him meant first persuading the German astrophysicist to get his legal papers entered into the public record. He enlisted the aid of Baade's best friend Milton Humason to try and talk the German expatriate into getting his affairs in order just in case. Milt and Baade went for a drive down to Mount Palomar one summer afternoon in 1938 to inspect the progress on the 200-inch dome. While on the road they talked about the changes that would be taking place at the observatory in the coming years. Adams was planning to retire once the 200-inch telescope was operational, and Baade was sure to be elevated to director at Mount Wilson. With the departure of Frederick Seares he would be in sole control of the photometry department as well. At the time everyone (but Adams) thought Hubble would become director at Palomar. Baade listened attentively to his friend and was grateful for the consultation.

In spite of Milt's best efforts, Baade remained intent on returning home to German soil. Not only did he love his country, but the directorship at Hamburg Observatory came with a salary of $7500 per month, well above the $3000 he then made at Mount Wilson. After moving some money around, Adams said he would be able to offer Baade a salary of $6000 monthly. The offer, while much lower than the one from Hamburg, was improved by the fact that the instruments and seeing conditions as well as the climate at Mount Wilson and Mount Palomar were far superior to those at Hamburg. Baade knew this and seriously considered changing his mind, but still procrastinated on the issue of getting his legal papers. In the end Baade's heart just wasn't in it (Fig. 9.2).

In the frenzy to secure the borders after the war began, the U.S. government set up a Provost Marshal's office in Los Angeles to manage, among other things, the removal of enemy aliens from the area. German and Japanese citizens and their children were required to register at the office and await orders. The German consulate in Los Angeles was closed for the duration of the war, and Germans living and working in the country were restricted in their movements. Even worse, because the staff of the observatory was engaged in optical work for the war effort,

Fig. 9.2 "Mile-a-minute Milt" and his new bicycle on his 50th birthday, 1941

Baade was prohibited from entering the grounds of the lab and from carrying on relations with any of the staff directly involved in the development of arms and accessories.

The final blow came in April of 1942 when a military curfew was established for "enemy aliens," requiring that they remain in their homes between the hours of 8 p.m. and 6 a.m. Stranded in a foreign country, without access to his country's embassy and unable to reach his family and friends at home, Baade would now be unable even to carry out his research. Baade had been grounded completely, and, even for the patient German astronomer, this development was too much. Baade pleaded with Adams for help.

Adams immediately sent a letter to Vannevar Bush, president of the Carnegie Institution of Washington, and head of the Office of Scientific Research and Development. Bush replied and said he would give what support he could.

Knowing Baade couldn't wait for word from Washington and wanting to show his unwavering support for the coveted German astrophysicist, Adams decided to try another approach. Milt and Helen had moved down the street from the Adams in 1940, so Adam's walked down to Milt's place for a talk. If Milt could somehow convince the Provost Marshal that Baade meant no harm, they just might be able to free up enough time for their friend and colleague to return to work. As for getting in the marshal's good graces, Milt could choose whatever means he thought necessary, with Adam's full backing. Milt agreed and phoned his friend to set up a time to meet at his house.

The next day, Milt drove to Baade's home on Foothill Boulevard and stood by the front door while his gimpy friend lumbered to the door and opened it. Baade's wife Johanna served tea as the two friends sat for a while talking through their plan to get their stories straight before heading down to the Provost Marshal. The colonel in charge knew Humason by name and asked him questions about the work going on at the observatory. Milt, as charming as ever, described some of the work he and Hubble had been working on, and invited the colonel to the mountain for an evening of stargazing. The colonel was duly impressed by Milt's country boy charm and listened intently as Milt explained how his absent-minded friend misplaced his papers and then summarily forgot the issue with all the work being done advancing the scientific developments at Mount Wilson. Baade happily swallowed his pride and let Milt do the talking. It took all of Milt's charm and guile to humor the colonel into acquiescing, but eventually the plan worked. After some discussion about the hours required for his study an exception was drawn up permitting Baade to be on the mountain during regular observing periods as part of his work. His movements would be strictly monitored, and he was only allowed access to the telescopes and the darkroom on the mountain.

With his friend's freedom to work secured, Milt's attention again turned to work on the war effort. Nearly everyone he knew was working trying to help the country win the war. On April 11, his cousin, David Witmer, had become the new chief architect of the Pentagon, which was rising slowly out of the ground near the Potomac River in Washington, D.C. Hubble left in the summer of 1942 for the Aberdeen Proving Ground, where he would spend the remainder of the war as head of exterior ballistics. From the onset of the war through 1946, Milt served as investigator and property officer for the Carnegie Institution of Washington, consulting on optical instruments at M.I.T. and aerial photography and bomber formations at Wright Field near Dayton, Ohio. A pilot training ground had been established at the airfield, which was named for Orville and Wilbur Wright, in 1910. In 1927 the facility had been transformed into a research and development center for the air corps.

On March 8, 1942, Ruth gave birth to a daughter, Ann, and the family rejoiced in the birth of the newest member of the Humason clan. Mother and child were reported to be in fine health, and the infant's grandparents were nearby to lend whatever support they could offer the young couple. The work caring for wounded coming back from the war meant that Ruth must leave her newborn at home for periods of time to work at the army hospital, and Billy enlisted Helen in caring for the little one while Ruth attended the sick and wounded.

The glow of sunshine from the birth of the little blessing didn't last long, however. In April, Milt's mother, Laura Humason, died suddenly at her home on Orchard Avenue in Pasadena. She was buried at nearby Mt. View Mausoleum, and Milt spent some time caring for his father, William, now widowed after almost 53 years of marriage. The death of his mother came as a shock to Milt and the family, who had been celebrating the birth of baby Ann for only a month.

There is a saying that tragedies come in threes. In October, while working in the hospital, Ruth Humason contracted tuberculosis and fell deathly ill. Milt rushed back from his war work to give support to his terrified son. With his wife in quarantine and the need to be by her side, Billy asked Milt and Helen to take over

Fig. 9.3 Milt with his
granddaughter, Ann Humason

care of young Ann. The reports from the hospital were grim. The disease had shut
down one of Ruth's lungs and was threatening to spread to other parts of her body.
Milt and Helen took Ann into their home, telling Billy they would care for her for
as long as necessary. If she was going to survive, Ruth's ordeal would be long and
difficult. With Billy keeping vigil the family hoped for the best but prepared for the
worst (Fig. 9.3).

Meanwhile, there was still the matter of Ginny in Holland and Milt's duties for
the war department as well as his usual work at the observatory, which was ongoing
if slowed considerably by events. From the beginning of the U.S. involvement to
the Japanese surrender in 1941, Milt managed to publish six papers on a variety of
subjects.

The news from the Pacific was grim. The Japanese had captured the Philippines,
forcing commanding officer General McArthur and members of the Philippine
government to flee the island nation under cover of darkness. Most of Europe and
Scandinavia as well as northern Africa were occupied by Axis forces, and all of the
pressure to rid the world of tyranny had fallen to England and the United States. By
the end of 1942, however, the Axis' advance began to falter bringing spirits up at
home and increasing the call of fellow citizens to do what they could to aid in the
effort to defeat the German and Japanese war machines.

By the fall of 1943 the Allies were beginning to turn the tide of the war, and the mood was more hopeful for total victory abroad. In October, the Humasons were dealt another tragedy. William Humason died when the boiler he had been working on exploded at his home. William was buried next to his wife at the cemetery the family had visited just the year before. Despondent, Milt again wrote Ginny with the bad news.

By this point in the war the loss of life abroad, the death of innocents, personal tragedy at home and Ginny's ordeal were beginning to take a toll. As a member of the scientific community, Milt was clued into some of the most top secret initiatives of the day. Colleagues working on the highly classified Manhattan Project in the desert near Santa Fe, New Mexico, told him an effort was under way to build an atomic bomb. It was feared the Germans were working toward the development of a similar weapon, and the race was on to be the first to test one. If the Germans got the bomb first, all would be lost. Franklin Roosevelt, at the behest of Albert Einstein, among others, had been committed to the development of the bomb almost from the onset of the war.

By the middle of 1945 Hitler had committed suicide and Germany had surrendered, leaving only the Japanese to finally subdue. Roosevelt had also died, having just been elected to an unprecedented fourth consecutive term in office, and Vice-President Harry Truman had succeeded him. Truman, like Roosevelt, was determined to use the atomic bomb should one become available. With the Japanese emperor Hirohito taking a no surrender posture, Truman reasoned that the strategic use of the bomb would ultimately save more lives than it took.

On the afternoon of July 15, the day before the Trinity test, Milt made one of his periodic drives down to Mount Palomar for an evening of observing. On his way up the winding road to the summit he stopped by a group of Palo Indians who were selling woven baskets and other trinkets outside their teepee on the mountain. After talking with them for a few minutes Milt bought an item or two to bring home to Helen and young Ann and then drove to the summit. The sky was clear that day, and it looked like he could count on a good night of observing. As it was on Mount Wilson, the seeing on Mount Palomar had been tested by Milt, Baade and others, and the mountain was deemed to be an ideal location for the site of the biggest telescope ever constructed. Other locations had been scouted in Chile and elsewhere, but in the end the decision was made to keep the giant new reflector and its observatory closer to home.

Rising into the sunset sky, the 200-inch dome stood like a great white domed cathedral above the high desert hillside. The surrounding hills faded into the skyline as Milt set up a 4-inch telescope for an evening of observing. In December of 1941, Milt had published a leaflet in the journal of the Astronomical Society of the Pacific. The article discussed the latest advancements in photography, opening with a depiction of the great so-called aerial telescopes used by astronomers such as Cassini and Huygens in the seventeenth century. Peering through the simple lens set on a pivot point high on a perch that was mounted on a tall pole, these wondrously ingenious creations had a focal length of some 200 feet and were used to solve mysteries such as the nature of Saturn's rings and the diameter of Venus. Advances in lens technology soon made it possible to build telescopes of much smaller focal length with much more light-gathering capability, and instruments of increasing aperture were built throughout the next two centuries. Milt then pointed out that, "owing to the present emergency," alluding to the war, the 200-inch would likely remain the largest telescope in the world for some time but added that other less expensive and much simpler advancements as those being made in optics were making major improvements in photography. Using new red-sensitive plates the 100-inch telescope more than surpassed the gain in speed expected in the 200-inch telescope at its inception in 1928 and would lead to a further increase in speed for the new telescope upon its completion.

As the sky darkened Milt began recording the seeing conditions. All night he stayed there, tracking several objects across the heavens until the dull glow of the predawn Sun slowly brightened the eastern sky. At 5:30 a.m. Milt turned to the East where 800 miles away, in the New Mexico desert, a great flash lit up the morning sky with the light of twelve suns as the first atomic test exploded at White Sands Proving Ground near Albuquerque. The Trinity device had the energy of 20 kT of TNT, leaving a crater 10 feet deep and 1100 feet wide. The shock wave was felt 100 miles away, and the mushroom cloud billowed 7 miles into the sky.

Standing in the arid stillness of that southern California morning Milt gazed into the uncertain distance and wondered. The Great War that had taken so many lives and unleashed unspeakable cruelty on innocents around the world was nearly at an end. What menace now lay in its wake? How strong was humankind's will to power? What would that power mean for the future? Lingering for a few moments more, Milt shuffled across the gritty path to the car and headed down the long mountain road for home.

Less than a month later, in separate bombing runs, atomic bombs were detonated over the Japanese cities of Hiroshima and Nagasaki, killing more than 100,000 people instantly and wounding tens of thousands more. In a few days the Japanese emperor surrendered, ending the war.

Chapter 10
Beginning Again

Abstract In 1946, Ruth Humason is reunited with her daughter after a three year battle with the disease that nearly killed her. The Hale Telescope is dedicated in a ceremony on Mt Palomar. A weary Edwin Hubble returns from war work in Maryland and tries to get back to his observing programs.

Her fourth birthday was an especially auspicious time for young Ann Humason. With her mother locked away in a hospital room trying to recover from the effects of tuberculosis Ann had grown up under the loving supervision of her grandparents. Hummy and Danny, as she called Helen and Milt, were there for every moment of her young life. Bill Humason, the head of production at Proctor and Gamble, made frequent visits to see Ann at her grandparents' home on 1149 San Pasqual St. in Pasadena. The L-shaped ranch-style house had been the home of the recently deceased geneticist, Thomas Hunt Morgan. Morgan had won the Nobel Prize in Physiology or Medicine in 1933 for his discoveries relating chromosomes to heredity from his long study of fruit flies at Columbia University. Morgan discovered that genes supplied the basis of heredity, forming the modern science of genetics. The home had living quarters at each end and a shared kitchen at the center. A long porch ran the entire distance of the home in the backyard, which was groomed with palm trees and a small pool.

During visits with his daughter, Bill showed her pictures of her mother and told her about her mother's battle and assured her that she would okay soon so they could all be together again. Her grandparents kept her mother in Ann's thoughts as well, hoping against hope for Ruth's speedy recovery. In 1946, Ruth Humason finally won her battle against tuberculosis and was released from the hospital with a clean bill of health.

On March 6, Ann watched as her father's car came to a halt on the street outside her home. Climbing out of the car he walked around to the passenger side door and opened it. A carefully groomed woman stepped from the car and stood before her in a knee length dress. She looked every bit like the woman she had been hearing about all her life, and now she was standing before her smiling away tears of joy (Fig. 10.1).

R.L. Voller, *The Muleskinner and the Stars*,
Springer Biographies, DOI 10.1007/978-1-4939-2880-4_10

Fig. 10.1 Bill and Ruth (Petty) Humason and their daughter, Ann, sit by the fountain at Milt and Helen's home near Caltech, 1946

Her little girl stood before her in a cute pleated dress, with ribbons in her short brown hair that was parted down the middle. Young Ann smiled awkwardly and looked up at her mother, who fought against the fear and anxiety of having missed so much of her daughter's young life. Slowly and surely the two took their first awkward steps toward a wonderful bond between mother and daughter.

That beautiful sunny spring day, the Humason family celebrated Ann's birthday on the back porch, all together for the first time. Ann opened gifts on the table while her father and mother and her doting grandparents looked on. It was the beginning of a long transition into a life lived with her parents, but it was a beginning.

In April of 1948 the Humasons were blessed with another long awaited homecoming. With life returning to normal and longing to see her family again, Ginny made her way to Pasadena from Holland for a visit. The war had been evil and much of Europe was still in ruin as Allied governments divided Europe, Africa and the Middle East between them. Ginny told tales of unthinkable crimes against humanity and her struggle to save those lives that she could. The food stores and provisions had been helpful, but not all had gotten through. Somehow through the many close calls, she and her family had survived. Everyone was grateful for that. A party was held in her honor at Milt's and Helen's place. Lewis, divorced from Beatrice Mayberry, was living in San Diego with a woman named Gretchen, whom he planned to marry the following year. The hit of the party was young Ann, who was the spitting image of her mother and who marveled at her great aunt's table manners as Ginny lifted her plate and licked it clean of every last morsel. Milt could see the stress of the war years had aged his little sister beyond her years, but there were signs that her usually ebullient spirit was returning. Later the two of them visited their parent's graves and reminisced about their childhood. After many long hard years, it looked as though life would finally come back together again (Fig. 10.2).

* * * * * * * * * * * * * * * * * * **

On June 3 Milt and Helen pulled into a space in the parking lot of the Palomar Observatory and parked. Throngs of people were making their way to the dome of the giant reflector to witness its inauguration. Twenty years had passed since George Hale first conceived of the giant reflector that was being dedicated ten years after his death. More than 1000 people had gathered for the dedication. Those in attendance included Noble Prize laureates, scientists and movie stars, while photographers and news reporters captured the momentous occasion for a waiting public.

The large crowd sat awe struck under the massive yoke and tube as the ceremony started, marveling at the sheer size of the instrument. The mirror, made from a new material known as Pyrex, weighed 20 tons and required ten months in the annealing process before it could be loaded onto a flatbed railcar and shipped west

Fig. 10.2 Milt and Virginia during her trip to Los Angeles in 1948

from New York. It had taken more than 180,000 man-hours to grind more than 5 tons of glass from the glass blank to create the concave shape. The 55-foot tube was nestled in an enormous steel yoke resembling a horseshoe, the entire assembly and mirror weighing in at over 500 tons. The gears that smoothly guided the telescope were lubricated using forced oil, making it possible to operate the telescope with a motor no stronger than a common sewing machine. The surrounding dome weighed about 1000 tons and was synchronized with Earth's movement. The finished telescope increased the observing capabilities by eight times over the 100-inch telescope and could detect the light from a candle at 10,000 miles. The great instrument was named the Hale Telescope in honor of George Ellery Hale's contributions to the field of astronomy.

The ceremony was presided over by James Rathwell Page, Chairman of the Board of Trustees for the California Institute of Technology, and speeches were given by Raymond Fosdick, President of the Rockefeller Foundation, Palomar Observatory Council Chairman, Max Mason and Caltech president, Lee DuBridge. After a talk by the Carnegie president, Vannevar Bush, on the newly formed partnership between Mount Wilson and Palomar observatories under the supervision of Caltech, the new director of the observatories, Ira Bowen, described the technical aspects of the telescope.

Bowen, a physicist at Caltech, had been carefully chosen by the organizers of the two observatories to avoid a confrontation between the institutions' two biggest stars, Edwin Hubble and Walter Baade.

Left alone during the war with an unusually dark sky (city-wide brownouts significantly reduced light pollution) and plenty of observing time, Baade had made a fundamental breakthrough on stellar and universal evolution. With most of the staff (especially Humason) preoccupied with war work Baade had almost free reign with the world's largest telescope for years. Using this to his advantage Baade set out, using new red-sensitive plates, to target M31, and several of its companions in the hopes of resolving the light from the center of these galaxies into stars.

After carefully studying the resolved stars in these regions, Baade concluded that there were two populations of stars in them. Population I stars, like our Sun, were younger (0–10 billion years old) and circled in roughly concentric orbits around the galactic plane in the dusty arms of spiral galaxies. Population II stars, such as red supergiant Betelgeuse, were older (10–13 billion years old), highly luminous and could be found either in or above the galactic plane and in the bulge near the center of a galaxy.

Comparable to the advances in the field of cosmology that Hubble and Humason had introduced with their discovery of the expansion of the universe, Baade's new star populations opened new avenues in the field of stellar evolution. The discovery made Walter Baade a star in the world of astrophysics. Through the middle of the twentieth century, only Jan Oort (Oort discovered the galactic halo, provided the first evidence of the existence of dark matter, the direction and distance of the galactic center from Earth and predicted the existence of an extreme outer region of the Solar System where long range comets rotated around the Sun) rivaled Baade in stature among the world's astronomers.

Edwin Hubble returned from his war work a changed man. A rock-ribbed conservative and outspoken supporter of Allied efforts in both world wars and a veteran of the first, Hubble was suddenly sullen and extroverted on the issue of nuclear weapons. Like his friend and Caltech professor J. Robert Oppenheimer, who was tormented by his role in the bombing of the Japanese cities of Hiroshima and Nagasaki, Hubble began a campaign to abolish the weapons shortly after the war ended. His frustration over the course of the observatories after being slighted by the Carnegie Institution and Caltech board members left a further distaste in his mouth.

In the lead up to the dedication of the new facility on Palomar the prevailing thought was that each of the observatories would be led by a director. It was thought at the time (and Hubble assumed) that Hubble would take over as director at Palomar and Walter Baade would be chosen to lead the Mount Wilson Observatory. Not a supporter of Hubble for director of the facility, the soon-to-be retired Walter Adams submitted that Hubble's appointment would lead to an unfair advantage in the administration of telescope time on the new reflector. This advantage, whether real or perceived, should be avoided at all costs, lest the institution should lose Baade, one of its stars, in the process.

Meanwhile a melodrama was being played out on Mount Wilson as Hubble's name was being slandered to prevent him from becoming director. In a letter to Mayall, Milt describes the tenor of things writing to his friend, "Believe it or not, there are many people who dislike Hubble and in every way possible they are trying to prevent him from being named director." Later as apparently Hubble made attempts to restore his image among the Mount Wilson staff, a more upbeat Humason wrote, "Hubble is going about things in a different sort of way, and the future looks pretty bright for him as far as the 200-inch is concerned."

With Hubble smarting over being snubbed for the directorship of the observatories and Baade itching for equal share of telescope time on Palomar, the new director, Ira Bowen, had his hands full. Bowen was a brilliant scientist with a mild, patient demeanor who had a great capacity for sorting out contentious details. He was not an astronomer, and although he was respected by those who knew him, he was reluctant in the early going to confront the two most established members of the staff. Hubble's public fame extended well beyond pure science, while Baade was one of the top two most respected astrophysicists in the world. Rather than provoke either man, Bowen preferred instead to allow time to get to know everyone and build trust. Bowen was a cool customer, and he needed an equally cool hand to work as secretary of the new combined observatories. For this job he chose Milt, figuring that Humason's work with several members of the staff and his ability to get along with everyone including Baade, who was a close friend, would be a leavening tool.

As secretary, Milt's responsibilities included scheduling observing time on both Mount Wilson and Palomar as well as reading and responding to letters from the public and other administrative duties. The mail he received from the general public ranged from the intellectual to the absurd. In one case, he received a very well thought out idea to publish a map of the universe and local galaxy clusters then known. The idea came in a long paper with wonderfully detailed drawings and maps that were designed to help laypeople understand the complex nature of the cosmos. The idea had come from an engineer who had been working on the problem in his spare time. Milt took his proposal seriously and helped get the book published. In another letter an amateur astronomer mused about the Orion Nebula, saying that he had always suspected that heaven lay just beyond the cloud nestled in the region just under Orion's Belt. The man wrote wondering if Milt had seen it. Milt replied that he had seen many millions of light years beyond Orion and had yet to see any sign of anything other than space and star stuff.

Public correspondences like this were a welcome diversion from the often daunting task of scheduling observing time. This unenviable job had been the burden of Alfred Joy for years and was now being split between Milt, acting on behalf of the stellar department, and his good friend Seth Nicholson, who scheduled time for the solar department.

Hubble's galaxy classification and expansion problems necessarily required long exposure times into deeper and deeper regions of space. The success of the program and the scientific and public interest in the question of expansion led to its dominance in telescope time at Mount Wilson, and it was expected that this would carry through at Palomar once the 200-inch came online. Although the evenhanded Humason tried his best to mete out the remaining time as best he could among the staff, problems inevitably arose and people felt slighted. When a member of the staff became too unruly Milt soon learned to use psychology, telling his ungrateful colleague, "All right, starting next month the scheduling job is yours." This usually quieted the protester down.

If scheduling observing time was already an undesirable task, trying to navigate the growing turf battle between Hubble and Baade was utterly thankless. Although he deeply respected both Hubble and Baade for their accomplishments, Milt ultimately leaned in the direction of Hubble's program, largely because the expansion problem required going deeper into space for evidence as well as candidates for the galaxy classification system. Having failed to reason with his friend Baade tried the next best thing. He asked Milt to make photographs of various objects for his population program during Milt's observing runs with the big telescopes. Busy with Hubble's program Milt couldn't make as much time for Baade's work as he or Baade would have liked, and the pace of data collection was very slow.

Thirsting for support for his work on stellar populations, Baade next reached out to Nick Mayall at the Lick Observatory for help. Although Mayall was already working with Hubble and Humason on the nebular classification program the young astronomer was more than happy to give time to Baade's work on stellar populations. Increasingly Baade's feeling that the administration at the observatory was engaged in favoritism would drive a wedge between the longtime friends. Just after Hubble's return in March of 1946, the two had an argument over observing time for Mayall, whom Baade had hoped to have down from the Lick to photograph emission nebulae in a follow up to a paper he'd published on possible evolutionary aspects in them. Humason, intent on making sure Mayall's work was carried through, intended to take spectra of various galaxies to test Mayall's earlier assertions. Baade thought Mayall should be given time to do his own work and was incensed by Humason's refusal to give it to him. In a letter to Mayall Baade made his frustration felt at what he perceived as Humason's selfishness, sarcastically referring to him as "the former mule driver," and calling him a "perfect conceited ass!" Milt published the paper "Frequency of Emission Lines in Extragalactic Nebulae As A Function of Nebular Type" in 1947, citing Mayall's previous work while adding the Mount Wilson data. The paper noted a marked increase in emission lines in spiral over elliptical galaxy types, indicating an influx of Type I population stars in late spiral galaxies, and alluded to the possible conclusion that these may represent an evolution from one stage of galactic development to another. The use of star populations in the paper was no doubt a central factor in Baade's ardent support of Mayall's work. Eager to work with any of the big three at Mount Wilson, Mayall avoided the conflict between Humason and Baade.

Problems with the Hale Telescope early on caused as much turmoil as it settled, at least temporarily. As was the case with any new instrument adjustments needed to be made to get the telescope working to its maximum capacity. This put to rest the issue of scheduling on Palomar for the moment but increased the stress on Milt as he tried to schedule Baade and the other astronomers on staff time with the 100- and 60-inch telescopes on Mount Wilson and the 48-inch Schmidt on Palomar. Citing the lack of one of the observatory's main instruments gave Milt a reprieve from scorn and a chance to attend to other issues and concerns.

In 1946, as the fog of war subsided, the staff was steadily returning to scientific work. On an observing run at the Cassegrain focus of the 100-inch, Alfred Joy called to the night assistant and asked him to take over for a while. The 64-year-old astronomer needed a break and closed the shutter at the telescope so the change could be made. As the two men passed each other in the same aerial dance they had performed countless times before, Joy lost his balance and plummeted 30 feet to the observing platform. Not on the mountain at the time, Milt heard the news the next day. Somehow, the old man of the mountain had survived, but he was badly injured in the fall, having broken both arms and legs. Milt was among those who helped Joy during his long recovery, stepping into take exposures of star regions for his long list of scientific studies. Joy eventually went back to work at the telescopes and continued to work after his retirement in 1948 until his death in 1973 at age 91.

Milt and Joy published a paper on a dwarf star that was discovered to be a companion of a nearby bright star at a distance of about one-second of arc. The discovery was another in a litany of so-called binary star systems that would be found in time to make up more than half of all star systems, the term binary having first been used by William Herschel in 1802. The paper was published in the Astronomical Society of the Pacific journal in June of 1949.

Meanwhile the greater weight on Humason's mind concerned the health of his partner on the expansion principle, Edwin Hubble, who was struggling to return to form since coming back from the East Coast after the war. Hubble had tried to pour himself back into photometric work, but technical delays at the 200-inch were preventing Humason from resuming his pursuit of higher velocities. The war had robbed both of five good years of research and delayed the completion of the 200-inch telescope by ten years. Now in their late fifties each was reaching the twilight of his career. Hubble was in far worse condition than his slightly younger collaborator.

New theories were increasing public interest in the work ongoing at Mount Wilson and Palomar. In 1949, a group of scientists and engineers from England led by Fred Hoyle introduced a new model for the universe called the Steady State theory. Hoyle, an astronomer, and his two counterparts, Thomas Gold, an engineer, and Hermann Bondi, a mathematician, put forth the idea that the universe was both expanding and unchanging. Hoyle rejected Lemaitre's version of the universe because it implied the existence of a creator. In two papers, one published by Gold and Bondi, the other by Hoyle, the trio discussed their idea in philosophical and mathematical terms. If the universe was infinite, they argued, then it could theoretically double in size and remain unchanged to the observer. To explain the formation of matter, Hoyle invented what he called a creation of a C-field, an invisible matter-producing sublayer that permeated the entire universe, producing the requisite one atom of matter per century per volume of space roughly equal to the Empire State Building.

The new theory became regarded as a plausible alternative to the Big Bang. Ironically, it was Hoyle who gave Lemaitre's theory its catchy name. In a series of lectures for the BBC Hoyle explained the differences between his Steady State theory and its competitor. A rebel at heart, Hoyle once likened Darwin's theory that life on Earth had evolved from simple matter was like sending a tornado through a junk yard and spontaneously creating a new jetliner. Hoyle noted that the proposed expansion caused by the initial explosion, as he called it, meant the universe we see today is a relic of that event. He disagreed with Lemaitre's conclusion on philosophical grounds, concluding, "I cannot see any good reason for preferring the Big Bang idea."

One argument against Lemaitre's theory stemmed from Hubble's law. Comparisons between Hubble's proposed distance-velocity scale of 500 km/s per megaparsec and the fossil records had led scientists to conclude that for Hubble's figure to be correct the universe would have to be younger than Earth by more than a billion years. Although the news was unsettling, Hubble and Humason were convinced that their velocities were correct within a very small degree of error and that whatever the fate or origin of the universe, its apparent uniformed expansion was real. They decided the best way forward was to take the velocities out to the limits of the Hale Telescope when it came online and ensure the relationship held up to those extremes. The 200-inch would likely be the last increase in aperture for twenty years and should be more than sufficient to resolve the issue of expansion as far as Hubble's law was concerned, if not the question of origins. In a moment of conjecture Hubble even conceived that the 200-inch might take them to the visible horizon of the known universe, the farthest distance from which light traveling through space would reach Earth.

As the 200-inch was finally put into production Milt began fishing for spectra of deep space galaxies. In May a first year student in astronomy at Caltech named Allan Sandage joined the staff on Santa Barbara Street as part of the graduate program. A naval veteran of the war Sandage had received his Master's Degree from the University of Illinois the year before and was studying with Walter Baade at the large reflectors on Mount Wilson. Sandage had read Hubble's book, *The Realm of the Nebulae,* and was eager to meet and work with one of his heroes in astronomy. Sandage had dreamed of working at the observatories ever since his first visit to the area in 1941 with his father, who was teaching summer school at Berkeley at the time.

Born in Ohio in 1926, Sandage had an inner drive to understand the correlation between the "the worldly and the otherworldly." He got his first look through a telescope as a boy and knew from that moment on that he wanted to become an astronomer. He counted Hubble among his heroes along with Galileo, Newton and Einstein, and relished the brief period he got to spend working with Hubble. Under the tutelage of Baade and Humason, Sandage had quickly learned the ropes of the different instruments on the two mountains and proved to be excellent at making photometric measures of direct photographs as well. So, during winter months for a few years starting in 1950, "Sandage for Hubble" was often the observer name given in the log book on Mount Palomar. Although Hubble had aged considerably and walked with a cane the meeting captivated the young astronomer. Hubble was impressed by Sandage's war record, his intelligence and eagerness to learn all he could about his chosen field. Shortly thereafter Sandage began working with Hubble, making measurements of star regions from plates the aging astronomer had given him.

In early July, Hubble and his wife boarded a train for Colorado on a month's long vacation. Approaching his 60th birthday, Hubble hoped that some time fishing and hiking in the mountain air would restore his health and energy. Humason was heading down to Palomar to take new spectra, and he hoped to return to nebular classification work on his return. Shortly after his arrival in Grand Junction Hubble suffered a massive heart attack that damaged one of the muscles in his heart and left him bedridden for months. His wife Grace wired Milt with the bad news. His condition prevented Hubble from being moved, and they would have to remain in Colorado for some time. By September he was home again, having ridden the rails back to Pasadena in a hospital car. Hubble's sudden collapse brought the sometimes combative members of the observatories together in support of one of their own. Walter Baade expressed his concern for his American counterpart saying, "Let's hope he pulls through."

With Hubble's condition uncertain, Milt launched himself into the nebular work as the big telescope on Mount Palomar came online late in 1949. Things were noticeably different at Palomar. The telescope, dome, the electronics, even the light switches all had a more modern feel to them. The massive tube faced up toward the sky, waiting for him as Milt climbed into a new heated suit used by pilots during the war and now available to astronomers on both mountains for cool evening runs on the telescope. Operating the telescope was different, too. Climbing into a small basket-like elevator Milt rode the rails up the side of the dome to a catwalk that stretched out over the opening in the tube. He stepped onto the catwalk and walked over to the end of the platform, climbing down into the tube. Seated at the end of the tube, warm in his military overalls, Milt guided the telescope out over the observatory floor to begin his run.

Hubble showed up at the office in Pasadena in October, but he was too weak to resume steady work and stayed at home most of the time. For the next year, while Humason continued to take spectra of galaxies, Hubble continued to recover, spending some time on the East Coast in 1950.

Arriving at his office in May Milt found a letter addressed to him from Knut Lundmark at the University of Lund in Sweden. Milt and Helen had met Lundmark and his wife and visited the university on a trip to Europe in 1938. The letter was addressed to "The Astronomer Dr. Milton Humason." In a warm tone the letter began, "My dear Milton." Lundmark was writing to inform Milt that he was to be awarded an Honorary Ph.D. from the university. He offered congratulations and hoped that Milt and Helen could make it out for the ceremony. Milt's reply mirrored Lundmark's tone, conveying his respect for the university and thanking the observatory director for his support.

Ginny wrote with beaming pride to send her congratulations on the eve of Milt's 59th birthday in August. An honorary degree, she wrote, "is the highest recognition there is for what one is, because it is offered in appreciation of what one has accomplished," and that it, "certifies that one has…done something important to a degree far above the ordinary." Perhaps understanding her brother's humble nature she continued, "As to what this all means to you…no one can ever know that except yourself." Ruminating further on the achievement she added, "someone…like myself, does have some dim idea of the long trail stretching between the boy at Strain's Camp…thus the 'mule-driver,' and the doctor of philosophy of today. I think that few doctors *honoris causa* will have climbed such a steep and rocky trail to reach that high peak…the view from the top must spread itself in extra splendor for you." The road had been steep and rugged, she continued in glowing praise, and added that no "railroad tycoon" stood as rich as Milt at this moment in his life. "*Per aspera ad astra,*" she added, "through adversity, to the stars." Kidding with him, Ginny ended the letter by noting that his birthday would be the first he would celebrate as Dr. Humason, a fact that she knew he would deny.

In spite of Lundmark's urgings, Milt did not go to Sweden to collect his award. His duties at the observatories and the ongoing work on nebular and other research were too much to allow him to leave the country for any extended period of time. He filled out the enclosed typewritten forms with biographical details and included a bibliography. One of the questions on the list asked if he was available for public speaking engagements, to which Humason typed an emphatic "Definitely no!" His discomfort in public settings was palpable even after years of experience and having become one of the leading experts in his field. Although he downplayed his contribution, Humason's humility was encroached by more honors. The Royal Astronomical Society made Milt a member in 1951.

Hubble was finally cleared for work on the mountain in October. Grace insisted that he be accompanied by Milt and Allan Sandage, who had become familiar to the Hubble family, visiting Hubble frequently at his home to work on photometric measurements for Hubble's program. Observing runs were restricted further by time of year. He was not allowed to observe during winter months and was filled in for in that capacity by either Humason or Sandage. During several observing runs, Milt had shown Sandage the best operating practices for the telescope, having worked through some of the bugs in its system. Sandage had by that time become Hubble's assistant and was present whenever Hubble was on Mount Palomar or the observatory offices in Pasadena.

By the end of 1950 Humason had velocities for six new galaxies. With exposure times of less than six hours he had extended the velocity-distance relation from 250 million light-years, the last estimated distance at Mount Wilson, to 350 million light-years. The fastest velocity was 61,000 km/s, about one-fifth the velocity of light. Humason included his findings in a note to the Los Angeles meeting of the Astronomical Society of the Pacific and later published it in the *ASP* journal in August 1951. The paper was Humason's 80th published work. The two-page note, entitled "Apparent Velocities of Extragalactic Nebulae in Four Faint Clusters," laid out the range of new velocities and touched on the change in distance and velocity from Mount Wilson to Palomar. Concluding the note Humason added that considering the relative ease in obtaining the first velocities with the 200-inch telescope, "an accurate formulation of the law of red-shifts can be established out to about twice the distance previously attained." It was the last paper Humason would publish for three years.

About the time his new velocities were being published Humason began having trouble on Palomar. Try as he might he could not get a definitive spectrum for any galaxy beyond 350 million light-years. He checked the telescope and spectrograph and made some adjustments before trying again with no better result. The H and K lines he was trying to expose on the red-sensitive plate were not visible. To his astonishment and frustration Humason had gone as far as he could go. Discussions were held and the staff engineers were brought in from the Carnegie and Caltech labs. It was finally discovered that the incredible light-gathering capability of the 200-inch mirror was absorbing the light pollution from Los Angeles and washing out the spectrum on the plates. Progress had interrupted progress.

Until a suitable solution to the light pollution problem could be found, the pursuit of spectra for galaxies beyond 350 million light-years had to be put on hold. There was still plenty of work to be done on the classification program, however, which included velocities to galaxies within the reach of the new telescope.

The next year, Walter Baade made an advance that bolstered the case for proponents of the Big Bang and dampened the spirits of Fred Hoyle and the Steady State supporters. Revisiting the period-luminosity equations for Cepheid variables Baade found the cosmic yardsticks fit into one or the other of his two stellar populations. Population I Cepheids, Baade discovered, were four times more luminous than their Population II cousins. The Andromeda Galaxy (M31) was filled mainly with Population I Cepheids, but the period-luminosity scale was built on

Population II Cepheids. Realizing that Hubble had made his initial estimates of the distance to M31 based on the less luminous Cepheids Baade revised the luminosity scale and in turn determined that the distance to M31 must be twice as far as was previously held, about 2 million light-years. The distance to M31 had been used in estimating the distance to other galaxies, so these distances, Baade reasoned, must also be doubled. If Humason's velocities were accurate, and he thought they were, then it was taking the galaxies he had been measuring twice as long to reach those distances, which meant that the universe must be twice as old as everyone thought. Before he was finished, Baade had doubled the age of the universe to 3.6 billion years. This meant that the universe was now older than Earth, a convenient fact for the supporters of the Big Bang theory. Baade announced his findings at the 1952 meeting of the International Astronomical Union in Rome to gasps of astonishment. Fred Hoyle was the official note taker at the meeting.

Just as disappointed as Hoyle, Edwin Hubble was devastated by the correction to his law of redshifts. Everyone around him thought Hubble was deserving of the Noble Prize for his contributions to science and Hubble agreed. It didn't matter that astronomy wasn't recognized at that point by the Nobel Committee. In view of the Baade's correction, Hubble resigned himself to the fact that he would not be honored with the prestigious award.

The work continued on galaxy classification, Humason and Sandage working at Mount Wilson and Palomar Observatories and Nick Mayall at the Lick. Humason and Mayall would take spectra for galaxies and Hubble or Sandage, in Hubble's stead, would take direct photographs of the areas for photometric measurements. Hubble's improving health was bolstering the spirits of his colleagues at both observatories.

Late in August of 1953, Hubble returned from a long trip to London, where he had delivered the George Darwin Lecture at the Royal Astronomical Society. He ended his lecture, "The Observational Evidence for an Expanding Universe," with a slide of the farthest galaxy yet taken at the 200-inch telescope proclaiming, "This is the last horizon." It had been a wonderful trip to his home away from home, which included a fishing trip with a friend in France where he addressed the French Academy. His energy restored and his spirits high Hubble left with Grace and drove to Palomar on September 1 for a few nights of observing. On Monday, September 28, 1953, Hubble was preparing for an observing run on Palomar and asked Milt to come by his office to discuss the course of research for the nebular classification program. The two old friends met for several minutes, and Humason noted his friends "ease and energy and how well he seemed," as Hubble said goodbye and began his walk home for lunch. It was the last time Humason saw him alive. On his walk home Hubble was picked up by Grace, who saw him walking by the side of the road, and the two headed towards home. Riding in the passenger seat of the car, Hubble suffered a stroke and died later that day in his bed at his home in Pasadena. No ceremony was held, there was no funeral and no burial plot for his friends to pay their final respects. It had always been Hubble's will to drift away quietly, and Grace made sure her beloved husband's final wish was granted.

Hubble's death came as a shock to everyone who knew him. For Humason, the loss of his friend and colleague with whom he had worked so closely for 25 years left a void that he would struggle to overcome. On behalf of his late friend, Humason resolved to finish the work they had started. He solicited the help Sandage and Mayall to complete his mission. Sandage would fill Hubble's role supplying photometric measurements, which he had been doing ably for several years, while Mayall would continue to take spectra for galaxies in the areas they had predetermined using the 36-inch Crossley reflector at the Lick Observatory. In Sandage and Mayall, Milt found the same resolve he carried to finish the task.

Work was slowed in the beginning as the three men worked with heavy hearts and somber spirits, and the universe seemed to conspire against them. That fall, Sandage left Pasadena for a meeting that slowed the pace still further. A forest fire on Mount Wilson in December preoccupied Milt's time as he worked with fire department and other local authorities trying to quell the inferno. In a letter to Sandage, he wrote his young friend with an update: "Sunday night it seemed as though the fire would sweep across Mount Wilson…Monday morning it…had moved up the canyon south of the monastery… It is still out of control, burning on the west slope of Mount Harvard and gradually working its way into Eaton Canyon. The most dangerous place for the observatory is the Echo Rock region. There it is within two to three hundred yards of the top of the mountain…. Some four hundred men are on Mount Wilson, including eleven big pumpers. We believe the instruments are safe, but there is still a possibility one or two of the cottages may be lost…. At this time, however, things look quite hopeful…. If ever a good soaking rain storm were needed, it is now".

Gradually the mood among the three astronomers brightened as they neared the finish line. Mayall spoke of being "fired up" about the coming compendium to the work of Hubble and Humason. They labored over every sentence of the paper, each trying to do his part to insure Hubble's legacy. As publication of the paper approached, Milt suggested that his younger counterparts take the top billing on the paper. Ever deferential Milt was probably trying to pass the torch to those would carry the effort forward that he and Hubble had so far advanced. In a letter to Sandage, Mayall stated his feelings on the matter: "…he suggests that we agree to putting his name last in the line-up of authors. My feelings are very strong that it should come first, in recognition of the really great amount of skillful, patient and reliable work he has done during many years in a difficult field. Next to Hubble, I regard him as setting a standard of real advance that will be hard to match or beat, in my time at least. From mule-skinner to Palomar's ruler of the redshift represents a rise having few counterparts in astronomy, unless it be a supernova!"

Sandage and Mayall were relishing the work assisting the aging legend in his final effort to complete the work he and Hubble had started in 1928. Letters between the two suggest they were sharing not only research but a little leisure time with Milt, joining him in his favorite pastime, fishing. At Sandage's urging, Milt sent a copy of the paper to H.P. Robertson for his input on their conclusions. Robertson was a master of Einstein relativity and was highly regarded for his interpretations of the theory as it pertained to cosmology and other problems, physical and theoretical. Einstein had died recently, and his sudden absence from the roster at the upcoming conference in Berne, Switzerland, left the committee with a void to fill, which they had hoped to do by inviting Milt as a keynote speaker on the subject of universal expansion. In a reply to Milt's letter asking his advice on whether he should attend, Robertson writes: "I think it would a fine thing if you or Allan would attend and give a condensed version of your excellent paper, as it gives the up-to-date word on the status of the observational results."

Humason didn't go to Berne, deciding instead to bury himself in the work on finishing the redshift project. A few days later Walter Baade stopped by for a visit and told Humason he planned to attack Hubble's approach to defining the principles guiding his law of redshifts. Baade referred to Hubble's magnitudes as "enthusiastic" and said his velocity ratio of 500 km/s per megaparsec was going to be corrected. Baade, who had spent years perfecting Frederick Seares' already very precise magnitude scales, was beginning to correct Hubble's velocity and intended to speak about his findings at the conference. In a letter to Robertson, who served on the committee at Berne, Sandage voiced his dissent on behalf of the three men, accusing Baade of harboring "extreme prejudice" at times and disagreeing with his former mentor's course of action: "Baade's comments last week were most disturbing. He indicated that he was going to discredit Hubble's work to 1936 on the precise evaluation of the constants and on his method of attack. We write this to you to let you know what may take place at the conference...Baade does carry the weight of a very distinguished observational astronomer. However, if he attempts to discredit Hubble or the observational basis of the redshift, Milt, Mayall and I do *not* agree with him."

Milt followed with a few lines written in support of Hubble and hoping his old friend would think twice about how he spoke about the issue:

> Dear Bob: I have read Allan's letter and want you to know that I am in complete agreement. Dr. Bowen also approves this way of handling it. We are sorry that Walter Baade feels that he must discredit Hubble's early work. It is not necessary, can not help, and seems a wrong thing for him to do. We do not object to his stating the case as he now believes it to be. We do resent what he has told us he is going to do at Berne in regard to Hubble's work. I feel almost certain that by the time he gets to Berne he will have calmed down considerably and will give one of his fine and interesting talks with the personal resentment angle out entirely. I have a great admiration for both Hubble and Baade and am sure they did for each other.

Taking Humason and Sandage's trepidations to heart Robertson replied with a short note: "I am very sorry to hear of Baade's curious behavior—I hope with Milt that he will get over it on the trip! I shall bear your and Milt's words in mind in Berne."

Their fears were dispelled in a letter from Robertson after the conference, in which he cites Baade's mood and lends his support for the German astronomer's reasoning in setting out his explanation for the correction in Hubble's velocity figure: "You and Milt will be glad to hear that Baade's report to the conference was quite restrained and, so far as I am acquainted with the situation, quite factual. It does seem to me quite in order in such a report to state the reasons why the previous distance scale must be revised, and this Baade did. You can't spring a factor of 3 on a bunch of physicists without telling them 'wha' hopponed!'"

Robertson had used the work of Hubble, Humason and Baade in his speech on the state of cosmological theory, wherein he briefly discussed the recent history of developments both theoretical and observational in altering the perception of the known universe for all time. He then alluded to the upcoming publication of the culmination of the Hubble program on redshifts and classifications of deep space galaxies in support of the current observational discoveries in the field of cosmology. Acknowledging Milt for lending him the results of the paper for the purpose, Robertson laid out the current corrections to the Hubble constant to "H = 180 km/s per megaparsec...implied by Baade's and Sandage's recent revisions of the nebular scale." With the trial of the conference behind them the team of Humason, Mayall and Sandage could relax again. In a letter to Sandage Mayall wishes them luck on an upcoming fishing trip, saying, "Hope you and Milt get those trout – and may they be big ones."

In April 1956, the *Astrophysical Journal* published a 74-page document entitled, "Redshifts and Magnitudes of Extragalactic Nebulae." This remarkable paper, the result of more than 25 years of work, 20 years in a combined effort with the Lick Observatory, represented the culmination of the Hubble program on redshifts. It stood as a testament to Hubble's intuitive brilliance and the incredible skill and perseverance of his chief collaborator, Humason. In keeping with the wishes of his current partners, Humason's name appeared first on the title line. The paper begins with a discussion of the history of the program to chart and classify the galaxies in the *Shapley-Ames Catalogue* according to type and to establish a relationship between velocity and distance, which Hubble had first announced in 1929. It summarizes the division of work between Mount Wilson, Palomar and Lick observatories in confirming the physical properties of over 800 galaxies from the *Shapley-Ames Catalogue* out to magnitude 12.9. In closing his chapter on nebular research, Humason had measured velocities for 620 galaxies, taking the velocity limit from 1800 km/s in 1929 to 61,000 km/s. Mayall had measured 300 galaxies at the Lick Observatory, and they had 114 galaxies in common, while Sandage stated the new known value for the Hubble constant at 180 km/s/mpc. Pages of charts, images of spectra and direct photographs supported the analysis of the data as Humason and team thoroughly and comprehensively discussed the data from various angles.

The results of the Hubble program set the tone for cosmology for the next fifty years and catapulted Sandage to the forefront of the field. With his commitment to his old friend and collaborator complete, Humason prepared himself for retirement. After 35 years of research and over 80 published papers on a range of astronomical problems, he had reached the mandatory retirement age. Although his retirement meant that he could no longer use the telescopes he was far from finished contributing to the search for answers to stellar and galactic evolution.

Chapter 11
An Aging Star

Abstract After the untimely death of his longtime friend and collaborator, Edwin Hubble, Humason takes up the task of finishing the program as far as the telescope on Mt Palomar can take it and publishing the work he and Hubble had started twenty-five years earlier. Soliciting the help of Nick Mayall and the Lick Observatory and Allan Sandage at the Carnegie Observatories, Milt and team publish the final results in 1956. With the work on the Hubble program complete Milt moves to finish his work with Fritz Zwicky on supernovas and the Hubble Atlas of Galaxies.

On May 11, 1956, the month after the final paper on Hubble's and Humason's long-term redshift and magnitude program was published, Walter Adams died at his home in Pasadena. Of the men and women who had touched Milt's professional career none was more influential than Adams. As champion, mentor, collaborator and friend, Adams had more to do with Milt's development in astronomy than any other. His fair hand and open-minded approach to managing the observatory paved the way for Milt's entry to the staff, and Milt could think of no one on the staff who didn't feel equally appreciative of Adams.

On December 20, 1954, the Carnegie observatories held a jubilee dinner hosted by Mrs. George Ellery Hale, to commemorate the 50th anniversary of the Carnegie Board of Trustee's decision to fund the Mount Wilson Observatory. As luck would have it the decision had come on Adams' birthday, and Hale must have felt his right hand man was a good luck charm that day. As replacements went, Hale couldn't have chosen better. Walter Adams' steady leadership and penetrating scientific mind had left the Hale legacy on firm footing. Milt and Helen remembered the trip they had taken with Adams and his wife to Camp Nelson in 1949 in Milt's brand-new Ford station wagon. Adams had retired in 1947, and the two couples had driven to the campground with young Ann for an afternoon of picnicking and fun. In an article he had published the year he retired on the early days at Mount Wilson, Adams had remembered the pack trains and people who were present during the early stages of the observatory's development. Milt was all that was left of that very early group of men who took the first bold steps up the mountain. Perhaps it was time to go.

© Springer Science+Business Media New York 2016
R.L. Voller, *The Muleskinner and the Stars*,
Springer Biographies, DOI 10.1007/978-1-4939-2880-4_11

But not if Fritz Zwicky had anything to say about it! Born in Bulgaria in 1898 to Swiss parents, Zwicky had studied at the Swiss Federal Institute of Technology in Zurich, earning a degree in physics in 1925 before emigrating to the United States, where he worked with Robert Millikan at Caltech. Brash and untrusting, the diminutive Zwicky was one of the most contentious characters at the observatory. He had a reputation of being difficult to work with and eventually managed to distance himself from nearly everyone except Humason. During his dispute with Baade in the late 1930s Zwicky referred to his one-time friend as a Nazi and called him a spherical bastard. When Milt asked him what he meant by that Zwicky replied, "No matter which way you look at him, he's still a bastard!"

The elfish Swiss astronomer was a bit of a wild man, capable of some outlandish ideas. He once suggested exploding bombs over Mount Palomar to clean the air for better seeing. His breathtaking trains of logic, rebelliousness and charisma made Zwicky as well liked among the student body at Caltech as he was disliked amid his peers.

When he wasn't taking theoretical leaps off the deep end Zwicky was capable of incredible and sometimes prescient insights. In 1933 he predicted the existence of dark matter, a ubiquitous layer of unseen matter that kept the rapidly expanding universe from pulling itself apart. Zwicky also predicted in 1937 that galaxy clusters could be used to bend light from a distant source around a massive foreground object to an observer on Earth in a technique he called gravitational lensing.

Another of these brilliant insights came in his 1934 paper with Walter Baade, when the two coined the phrase "supernova" to describe a particularly bright form of nova most or all of whom existed in galaxies outside our own. In an article on the subject from 1936, Humason describes the nova discovered by Tycho Brahe as being the only possible supernova candidate in the Milky Way. Tycho's supernova, which burst forth in November of 1572, "was brighter than Venus and could easily be seen in the daytime," according to Humason. Together Zwicky and Baade questioned whether these stars were the remnants of an explosive transition of an ordinary star to a neutron star, the small and densest stars in the known universe. They further suggested that supernovae might be the cause of cosmic rays. In another report that year on the spectrum of a supernova in NGC 4273 that had been discovered by Hubble and Glen Moore, Humason noted the existence of wide emission bands (200 A) "indicating that gaseous shells are expelled at great speeds," as both Baade and Zwicky had predicted.

In the search for supernovae that followed, Zwicky became the undisputed champion, collecting 120 of the bright exploded stars over his long career. In 1938, after his disputes with Baade had boiled over, and the two had stopped working together regularly, Zwicky recruited Milt to help him in the search and to promote the science. The veteran spectroscopist had put the nebular search on hold, having reached the limits of the 100-inch telescope, and had some time on his hands. Humason found a supernova of his own while on a visual search of the Whirlpool Galaxy (M51) in April of 1945.

Observing time having been shortened during the war, and with the resumption of the Hubble program in the late 1940s, Milt's assistance on the supernova program was for the most part impossible before 1957. In retirement he resumed his interest, and Zwicky and he formed a partnership in the search for supernovae. With his new partner at the ready, Zwicky launched a program of combined observatories in the United States, Switzerland and France in order to study them more closely.

The primary tools for the search at the Caltech observatories were the 48-inch Schmidt and 200-inch Hale telescopes. Using the 48-inch Schmidt telescope, whose wide-field lens made surveying large patches of the sky more possible than ever before, Humason and company tracked some 4000 galaxies in 64 fields two to five days a month all year long. The Schmidt telescope with its single axis and a guide scope easily accessible from the observing floor was much more civilized than the top of the tube of the 200-inch for a man approaching his seventieth birthday.

They began the search in 1958, and over the course of that year six supernovae were found in separate galaxies, four of them by Humason. Allan Sandage and Halton Arp made photoelectric measurements of the supernova candidates with the 200-inch while Milt and others made photometric plates. By 1960 the team had measured 19 more supernovae, Humason having found 14 of them independently and one with Paul Wild of the Berne Observatory in Switzerland.

By now the organization of the group was becoming clearer. Humason, the estimable astronomer, guided a small team of young researchers and research fellows in the operation and maintenance of the telescopes, the technical details of direct photography and spectroscopy at the instruments and the analysis and conclusions of the data from their observing runs. On the first night of an observing run, Milt would observe the field of galaxies in clusters to spot possible candidates and outline the scope of the field. Others, like Zwicky, would photograph spectra of the various candidates at the 200-inch for Humason, who would measure their velocities. The next year the group found 16 more of the ultra-bright exploded stars, Humason having found 11 of them. A report on the progress of knowledge in the field was published in October of 1961, led by Zwicky. The report pointed out that as of the publication of Hubble's book, *The Realm of the Nebulae,* in 1936, the word supernova wasn't being used in scientific nomenclature. During the period from 1941 to 1956, when no work was officially being done, an additional 54 supernovae were reported and added to the overall list of 96 discoveries in the first 30 years of research on the topic (Fig. 11.1).

The results of the present survey confirmed the earlier survey while providing clues about the course of research in the future. At their maximum brightness, they reported, supernovae were only one magnitude fainter than the brightest galaxies and occurred with the same frequency in most galaxy types, usually in the outlying areas. Zwicky proposed the possible use of certain types of supernova as distance markers but said explorations of more distant galaxies would be necessary to create a reliable formula. Spectra of the objects were not useful in determining their composition, but the team was confident that new techniques would reveal clues about their makeup in the future.

At the same time as they were undertaking the supernova search at Palomar, Zwicky and Humason started a systematic study of galaxies in clusters and multiple galaxies. The idea was to return to several of the clusters formerly measured for

Fig. 11.1 Milton Humason
around his 70th birthday in
June of 1961

their velocities for the distance-scale program of Hubble and Humason, to gather more information about the spectra of individual galaxies, especially peculiar ones. Among these were double and triple galaxies in collision with each other in a deep space intergalactic showdown. In a five-paper series the pair reported extensively on these frequently seen clusters and their strange double and triple constituents. Their findings were a starting point for further study by astronomers like Halton Arp whose 1966 book *Atlas of Peculiar Galaxies* beautifully illustrated the awesome nature of these odd congregations of stars.

At 70 years old, Humason was definitely enjoying his retirement. His granddaughter, Ann, got engaged in 1962, and Helen was helping her and her mother, Ruth, with the wedding arrangements. As the elder statesman of the nebular and spectroscopy departments, Milt's name was first on the list of new additions to the staff and visiting astronomers wanting to spend time with the living legend of Mount Wilson. Retirement hadn't dampened Milt's love of the stars, and he wanted to keep learning what he could about them. There were young astronomers to educate on the use of the telescopes and the craft of stellar photography. Humason delighted in the opportunity to teach someone a few tricks of the trade or regale them with stories of the observatory and the early days on the mountain.

An example of Milt's stature and his desire to bring along the next generation came in 1961 when, on an observing run with Schmidt telescope, he discovered a comet. The comet was given the name Humason, and Milt let two of the young members of the staff, Jesse Greenstein and Charles Kearns, who had been working closely with Milt on other projects, study the object as part of their research. The starry glint in the eyes of the aspiring astronomers reminded Milt of his first enthusiastic steps into the field with his good friend Seth Nicholson four decades earlier. Nicholson died July 2, 1963, having been bedridden for months in the hospital. That June Nicholson had been awarded the Catherine Bruce Gold Medal by the Astronomical Society of the Pacific in honor of his contributions to science. Confined to his hospital bed, Nicholson had to listen to the award ceremony over the telephone. Milt felt the sting of this loss as hard as any in his life. Milt's respect and admiration for his friend ran deep. Nicholson had been a patient mentor during his early years at the observatory. A highly respected scientist Nicholson was equally active in his community.

As he began his fifth decade of research Humason was finally reaching the end of the road. One more publication remained on his mind, however. In the mid-1950s Allan Sandage, with Humason's support, had begun the process of creating an atlas to commemorate Hubble's contribution to science. Any attempt by Sandage to include Milt's name with Hubble's or equate it with his discoveries was summarily quashed by the old spectroscopist, who saw the atlas as an ode to the late mariner of the nebulae. Sandage later had a chance to speak freely about his respect and admiration for Humason in his book, *The Centennial History of the Carnegie Institution of Washington, Volume 1: The Mount Wilson Observatory*. Ever the meticulous men of science, Sandage and Humason spent hours deliberating over the best printing process for the slides they planned to use in the atlas. They finally settled on the latest model for large-scale high resolution photographs from the Meriden Gravure Company in Meriden, Connecticut, around June of 1955. The following year, with the Hubble program on galaxies behind them, Milt and Sandy turned their focus again to finishing the manuscript for the atlas. They finally completed it in September of 1958 and mailed it to Mrs. Bauer at the Carnegie Institution for preproduction. It would be a full two years before the *Hubble Atlas of Galaxies* finally saw first light.

As Hubble's atlas was being readied for publication, Sandage began preparing a list of those who would receive complimentary copies. The list read like a veritable Who's Who in Astronomy of the era. He made sure to send one to Fred Hoyle, who was then at St. John's College in Cambridge and one of the staunchest detractors of the Big Bang. The debate between Hoyle's Steady State theory and the Big Bang was still an open question. No evidence of the point that the universe began had been found to support the Big Bang idea, and Hoyle and his supporters had yet to explain why no younger galaxies were visible in his version of the universe. Hoyle had established his reputation with his discovery of the means by which stars create the heavier elements during the different stages of stellar nucleosynthesis. This was a major breakthrough and led to further knowledge of stellar evolution.

The dispute between the Big Bang and Steady State theories was finally settled in 1964, when Arno Penzias and Robert Wilson, using a telescope at Bell Labs that had been converted from a radio antenna, accidentally discovered the cosmic microwave background radiation (CMB), a kind of cosmic echo created by the Big Bang explosion. In the Steady State theory, there was no bang, so there should be no echo. The discovery of the CMB won Penzias and Wilson a Nobel Prize and ended the debate between the Big Bang and Steady State. Wilson, who had studied with Hoyle and was a Steady Stater himself, would later comment that the breakthrough he and his partner had made was a bittersweet victory in that it disproved Hoyle's work, adding, "I very much liked the Steady State universe. Philosophically, I still sort of like it."

Another longtime Hubble nemesis, Harlow Shapley, sent a brief note to Milt on a postcard upon receiving his honorary copy, implying he would be too busy for a visit to the area that summer but saying he would be in Berkeley for a time. One can only imagine the what-ifs that must have been going through his mind upon seeing this volume for the first time.

After receiving the book, Nick Mayall commented that the use of the word 'galaxy' in a volume bearing Hubble's name was strange. Hubble preferred the term 'extragalactic nebula' and fought against the use of the other all his life. Galaxy was now the generally accepted term, while nebula was being used to describe an indistinct cloud of dust and gas.

Milt's copy arrived at his home on Poinsettia Street in Corona Del Mar on a sunny afternoon in June 1961. The beautifully illustrated book represented the third revision of Hubble's popular classification system for galaxies. In the Preface, Sandage discussed the reasoning, details and importance of finishing the work Hubble had started in the mid-1920s. The volume is still widely used by astronomers as a guide to the galaxies and their physical attributes. Nearly all the full-page photographs are from Humason and Hubble.

Epilogue

Retired to Northern California, Milt spends his remaining days fishing for steelhead. In an interview given at the time he gives his account of the moment he produced the enormous redshift back in 1928 and the work with Hubble on the Big Bang Theory.

In a 1965 interview a reporter caught up with Milt at his home in Mendocino, on the California coast at the mouth of the Big River. Ann had married and moved to Sebastopol, a little community north of San Francisco, to raise a family, and Milt and Helen wanted to be with their great grandchildren. Milt's old friend Hugo Benioff had moved there not long before and assured Milt the fish were biting and there was plenty of good weather. Fishing the rivers and bays around Mendocino County sounded perfect to the aging astronomer.

The Humasons were fond of hosting parties and had frequent visitors to the house on Mendocino Drive. Among those who signed their guest book were Milt's brother, Lewis Humason, and his new wife, Gretchen, Adeline Adams, the widow of Walter Adams, and Fritz Zwicky and his wife.

Life had slowed down considerably for Milt since he left the observatory for good. He still heard from Sandage and the others in Pasadena from time to time, and was always happy to hear from them and grateful to hear the latest on the events unfolding in astronomy. New technological advances were making life easier for the astronomers on both mountains. The mercury floats were being replaced by forced oil bearings that were much more dependable and created much less hazard than their highly toxic predecessor. Sandage had discovered quasars, quasi-stellar radio sources, at Palomar and was adjusting the Hubble constant to ever lower levels, expanding the known universe by billions of years.

His life in research had been replaced, however, and now Milt basked in the leisurely glow of a retirement well earned, fly-fishing the cool California waters while revisiting tales of the old days with his friend Hugo Benioff. The surrounding hillside, lush with pine forests and fields of wildflowers, made for wonderful picnic grounds for him and Helen to visit or take Ann and her family on their visits to the area. The hills east of the coast spilled into high bluffs that overlooked the sea. The Big River cut its way into the coastline, where salmon leaped from rapid to rapid in search of their spawning holes. It was an ideal setting.

© Springer Science+Business Media New York 2016
R.L. Voller, *The Muleskinner and the Stars*,
Springer Biographies, DOI 10.1007/978-1-4939-2880-4

The world had changed dramatically since his early days on the shores of the Mississippi River. In the seventy years since his birth the world had gone from the horse and buggy to the space race. In the early years of the Cold War, the Russian satellite Sputnik had been launched, and Yuri Gagarin had become the first man to orbit Earth from space in 1961. In a challenge to his country's scientific community, then President John F Kennedy had declared the U.S. intention of putting a man on the Moon in that decade.

Fritz Zwicky settled into a seat in Humason's home as the interviewer began to ask Milt about the program to measure redshifts with Hubble. Every now and again the life he had left behind in Pasadena would catch up to him for a moment. Milt talked about the history of nebular research, of Slipher's early redshift, about the type of glass that was used in the early spectrographs and of the advances made with the Rayton lens that had made it possible to take the first large redshift of NGC 7619. He had trouble describing his feelings about that first redshift, only saying that it made him "happy to see it." At the mention of Palomar Milt reluctantly got up to get his records so that he could bear witness to the evidence properly. Although he had written on the subject of expansion during the heyday of the nebular program he was reticent about lending his opinion on the subject in print. Asked for his thoughts on the subject, Humason demurred, saying only, "I have always been rather happy that my end of—my part in the work—was, you might say fundamental." When the subject of Hubble came up Humason struck a protectionist tone, stopping the tape and glossing over the 20-year history between the two carefully, so as not to provoke suspicion that might harm his friend's legacy. As the interviewer wound down he asked Milt if he missed working at the observatory. "No, no, no," he said. "I want to fish and that's what I've been doing ever since I'm up here. Steelhead and salmon, that's my business now." Like everything else in his life, Milt enjoyed his retirement to its fullest, fishing the rivers and waters of northern California until his death on June 18, 1972.

Milton La Salle Humason belongs to a unique group of men and women through the ages who pioneered in subtle ways to help broaden our view of the world around us. In a field where people worked tirelessly to promote their work and intellect, his quiet, homespun nature and sense of justice were irrepressibly charming. Although plagued by them earlier in his career, Humason learned to make peace with his insecurities, endlessly promoting those around him and never taking credit where he felt it wasn't warranted. If in the end more credit was due him than he accepted, those around him knew it and did their best to make sure the rest of the world did, too. Humason had the respect of some of the titans of astronomy—Adams, Merrill, Hubble, Sandage, Mayall, Baade and many others from around the world—who knew and understood implicitly his contribution to the advancement of science in their time.

What makes his story even more remarkable is the humble roots from which he rose, from a barefoot boy skipping stones in the Mississippi River in Winona, Minnesota, to the wide-eyed young muleskinner astride his horse on the narrow trails of Mount Wilson to Pasadena orange rancher, husband and father, to observatory janitor and on to his meteoric rise to the top of the astronomy world.

A simple, down to Earth and charismatic man, Humason had a reputation for being kind and amicable as well as mischievous and witty. He was by all accounts an excellent storyteller and could empty pockets around a poker table quicker than you can say, "Deal me in." These qualities and more are what later prompted Humason's protégé, friend and collaborator, Allan Sandage, to refer to him as "a superlative mule driver, fisherman, imprecationist, drinker, poker player, raconteur, rake and rogue, gentleman and friend."

As a scientist, Humason was slightly more complex than the trail-riding cowboy of his early days on Mount Wilson might suggest. Although ambivalent toward and fearful of seeking fame, he was, to some extent, tempted by its trappings. But Humason was, more than anything, guided and dazzled by the twinkling jewels in the sky, and in his pursuit to understand their mysteries, he became a most careful, meticulous, and ingenious technician. His pioneering work in a rugged and somewhat dangerous field (in the days before modern technology made observing simpler) stand in testament to his dedication and skill. Driven by curiosity and his love for his craft, Humason always strove to understand everything he could about the art and technology of stellar photography and worked with others to improve the instruments of the field.

The complexity in him only showed when the pale light of public interest was cast upon him. In these moments he found no structure from his own past to stand upon. Not one given to ostentatious displays, in these circumstances he simply demurred, and in the process shrouded his very worthy story in secrecy for years. Plainspoken and lacking the kind of formal education that might have cured some of his insecurities, Humason navigated through numerous interviews, public speaking engagements, photographs and calls of an interested public before ducking into the crowd.

He was even reluctant to get involved in print. When his cousin, Thomas A. Humason, Jr., an editor at Harcourt, Brace and Company, wrote Milt in October of 1950 (shortly after Humason had been awarded his honorary degree from the University of Lund), to ask him if he would like to write a book on the current state of astronomical discovery, Milt declined, saying, "It is true, as you say, that a book (to rank with the works of Jeans and Eddington) is needed now…It is certainly most kind of you to think of me in this connection—but I am not a writer!" He adds that, even if he did qualify for the job on a technical level, he lacked the literary skills to undertake the task.

This exchange is illustrative of all Humason's dealings with the public. Although he was more than qualified to write a book on the subject at the time when his cousin offered it to him, Humason viewed his literary ability as not being in step with his level of understanding of various subjects surrounding stellar, galactic and the universe's evolution. In this small way, Humason denied us a unique and singular voice in the field of science. A book on the stars written from the perspective of an old muleskinner turned astronomer would no doubt have been a delightful departure from the staid and stolid prose of the scientific community of his era.

The other argument to be made, however, is that Humason not only knew his place but was more than content to remain in it. In fact, he really enjoyed it, remaining at the helm of the giant reflectors on Mount Palomar and Mount Wilson until he was in his seventies, well into his retirement and mentoring others in the science that had given him so much. This sense of objectivity is a trait all too often taken for granted in the world today.

His personal attributes made Humason a friend to virtually everyone around him and a valuable partner and ally in the occasionally volatile world of competitive science. When combined with his skill and experience, Milt's character made him an excellent teacher. As his friend and collaborator, Nick Mayall, would later recall: "Milton Humason was the best mentor (at the observatory). He really knew nearly everything about each telescope, auxiliary instrument, and person on the mountain." All of these traits had been carefully learned throughout his life on Mount Wilson.

Whatever the causes or consequences of his character, there is little doubt of Humason's contribution to science. Probably the twentieth-century's most skilled stellar photographer, Humason revealed data on worlds never before seen and collaborated on some of the greatest discoveries of any age, chief among them universal expansion, to which he lent considerable evidence. As an administrator Humason was always aware of the needs of the staff and was known to be fair to virtually every man at the observatory, with the notable exception of election season. In scheduling observing time on Election Day, Humason quietly made sure the Democrats on the staff were on the mountain, rendering it impossible for them to make it to the polls to cast their votes. Noticing this Allan Sandage, who knew Milt to be a "rabid Republican," approached the aging secretary to ask him how such a coincidence could occur every Election Day. Humason was contrite on the matter, confessing to the slight without batting an eye. When Sandage once complained in outrage Humason informed him that if he didn't like it he should talk to Seth Nicholson, who had been doing the same thing to the Republicans on the solar staff for years!

As a scientist, Milton Humason lent his incredible skill and dedication to the discoveries of stellar and galactic evolution, helping to pave the way forward in our quest to understand the universe and setting a standard for generations of researchers. As a man, he was a loyal and highly entertaining friend, a devoted husband and that seemingly rare individual who was acutely aware of his place and always tried to treat those around him with respect. Fritz Zwicky once wrote in ode to Humason, "He always had the goal of a sound society and a beautiful world in mind," a statement that, if he had heard it, would have made Milt simply shrug his shoulders and deal the cards. Still, there is no discounting the grace and charm of the rugged Renaissance man of Mount Wilson, who climbed to the peak of the mountain and ventured hundreds of trillions of miles into the known universe. If one could imagine him having to write his own epitaph he would undoubtedly have kept it simple…

Gone fishing!

Bibliography

Osterbrock, Donald E., Pauper and Prince: Ritchey, Hale and Big American Telescopes; The University of Arizona Press, 1993.

Osterbrock, Donald E, Walter Baade: A Life in Astrophysics, Princeton University Press, 2001.

Hough, Susan Elizabeth, Richter's Scale: Measure of An Earthquake, Measure of A Man, Princeton University Press, 2007.

Read, Nat B., Don Benito Wilson: From Mountain Man to Mayor, Angel City Press, 2008.

Wright, Helen, A Biography of George Ellery Hale, E.P. Dutton and Company, 1966.

Clerke, Anges M., A Popular History of Astronomy During the Nineteenth Century, Forgotten Books, First Published in London by Adam and Charles Black, 1893?.

Brands, H.W., American Colossus: The Triumph of Capitalism, 1865-1900, Anchor Books, 2011.

Comer, Virginia Linden, In Victorian Los Angeles, The Witmers of Crown Hill, Talbot Press, 1988.

Johnson, Steven, How We Got to Now: Six Innovations That Made the Modern World, Riverhead Books, 2014.

King, Henry C., The History of the Telescope, Dover Press, 2003.

Deurbrouck, Jo, Stalked By A Mountain Lion, Falcon Guide, 2007.

Sandage, Allan, Centennial History of the Carnegie Institution of Washington, Vol 1 - The Mount Wilson Observatory, Cambridge University Press, 2004.

Carroll, Bradley W., and Ostlie, Dale A., An Introduction to Modern Astrophysics, Pearson - Addison Wesley, 2007.

Arp, Halton, Quasars, Redshifts and Controversies, Interstellar Media, 1987.

Gwynne, S.C., Empire of the Summer Moon, Simon and Schuster, 2010.

Sagan, Carl, Cosmos, Random House, 1980.

Drury, Bob and Clavin, Tom, The Heart of Everything That Is: The Untold Story of Red Cloud, An American Legend, Simon and Schuster, 2013.

Nasaw, David, Andrew Carnegie, Penguin Books, 2006.

Ambrose, Stephen E., Nothing Like It in the World: The Men Who Built the Transcontinental Railroad, 1863-1869, Simon and Schuster, 2000.

Brinkley, Alan and Dyer, Davis, The American Presidency, Haughton Mifflin, 2004.

Eisenhower, David, Going Home to Glory: A Memoir of Life with Dwight D. Eisenhower, 1961-1969.

Kazin, Michael, Encyclopedia of American Political History, Princeton University Press, 2011.

Larson, Erik, The Devil in the White City: Murder, Magic, and Madness at the Fair That Changed America, Vintage Books.

Lowe, David Garrard, The Great Chicago Fire: In Eyewitness Accounts and 70 Contemporary Photographs and Illustrations, Dover Publications, 1979.

Goldfield, David, America Aflame: How the Civil War Created A Nation, Bloomsbury Press.

Shulman, Seth, The Telephone Gambit: Chasing Alexander Graham Bell's Secret, Norton and Company, 2008.

© Springer Science+Business Media New York 2016

R.L. Voller, *The Muleskinner and the Stars*,

Springer Biographies, DOI 10.1007/978-1-4939-2880-4

Alef, Daniel, Cornelius Vanderbilt: The Colossus of Roads, Titans of Fortune Publishing, 2010.

LaPointe, Ernie, Sitting Bull: His Life and Legacy, Gibbs Smith, 2011.

Yenne, Bill, Indian Wars: The Campaign for the American West, Westholme Publishing, 2006.

Grant, Ulysses Simpson, Personal Memoires of U.S. Grant, Dover Publications, 1995.

Carnegie, Andrew, The Autobiography of Andrew Carnegie and His Essay The Gospel of Wealth, Penguin Group, 2006.

Christianson, Gale E, Edwin Hubble: Mariner of the Nebulae, The University of Chicago Press, 1995.

Singh, Simon, Big Bang: The Origin of the Universe, Fourth Estate, 2004.

Einstein, Albert, The Meaning of Relativity, Princeton University Press, 1922.

Johnson, George, Miss Leavitt's Stars, W. W. Norton and Company, 2006.

Osterbrock, Donald E. Eye on the Sky: Lick Observatory's First Century, University of California Press, 1988.

The Mount Wilson Contributions to the Carnegie Institution of Washington Yearbook, volumes 1-62.

Much of the study for this book was undertaken at the Huntington Library Munger Research Center. Additional study was done at the Lick Observatory library at the University of California at Santa Cruz as well as the Chicago, Winona, Los Angeles and Pasadena Public Libraries.

Index

© Springer Science+Business Media New York 2016
R.L. Voller, *The Muleskinner and the Stars*,
Springer Biographies, DOI 10.1007/978-1-4939-2880-4

Printed in the United States
By Bookmasters